Solid-State NMR and FT IR Studies on Chromatographic Column Materials

Von der Fakultät Chemie der Universität Stuttgart
zur Erlangung der Würde eines
Doktors der Naturwissenschaften (Dr. rer. nat.)
genehmigte Abhandlung

Vorgelegt von

Gokulakrishnan Srinivasan

aus Salem (Tamil Nadu), Indien

Hauptberichter:	Prof. Dr. Klaus Müller
Mitberichter:	Prof. Dr. Fritz Aldinger
Mitprüfer und Prüfungsvorsitzender:	Prof. Dr. Helmut Bertagnolli
Tag der mündlichen Prüfung:	24. August 2005

Institut für Physikalische Chemie der Universität Stuttgart

2005

Bibliografische Information Der Deutschen Bibliothek

Die Deutsche Bibliothek verzeichnet diese Publikation in der Deutschen
Nationalbibliografie; detaillierte bibliografische Daten sind im Internet über
http://dnb.ddb.de abrufbar.

ISBN 3-8325-1056-7

Logos Verlag Berlin
Comeniushof, Gubener Str. 47,
10243 Berlin
Tel.: +49 030 42 85 10 90
Fax: +49 030 42 85 10 92
INTERNET: http://www.logos-verlag.de

Dedicated to

my grandma (late) P.Vijayalakshmi, dad C. Srinivasan, mom S. Lalitha
and sisters S. Bhuvaneswari, S. Anandhi

Acknowledgment

I sincerely thank Prof. Dr. Klaus Müller for giving me the opportunity to pursue my doctoral degree under his guidance and for sharing his expertise, being positive and open-minded to all my suggestions and ideas. Words seem to be inadequate to express my heartfelt gratitude to him for his diligent guidance, constant help and suggestions throughout the period of my thesis.

I would like to thank Graduiertenkolleg Chemie in Interphasen for my doctoral fellowship and DFG (Deutsche Forschungsgemeinschaft) for financial support of this project.

I am grateful to all my co-authors and collaborators during the course of my thesis. I would like to thank Dr. Matthias Pursch, Dow Deutschland and Dr. Lane Sander, NIST for providing the C_{18} and C_{30} stationary phase materials. I would like to acknowledge Dr. Clayton McNeff, Zirchrom Separations Inc., and Dr. Angelos Kyrlidis, Cabot Corporation for providing carbon clad zirconia based stationary phases. I would like to acknowledge Prof. Dr. Helmut Bertagnolli, Institute of Physical Chemistry for allowing me to use the Raman Spectrometer and for the collaboration work. I also thank Prof. Dr. Klaus Albert and Dr. Hans-Joachim Egelhaaf, University of Tuebingen and Dr. Silvia Gross, CNR-ISTM, University of Padova for the collaboration work.

I am indebted to P. Anand Kumar, Dr. Karthikeyan Sharavanan, Dr. P. Muthukumar and my former colleague Dr. Shashikala Neumann-Singh for their great help during my search for a PhD position.

My hearty thanks to my former colleagues at the University of Stuttgart, Dr. Shashikala Neumann-Singh, Dr. Frank Berger, Dr. Thomas Handel, Dr. Tillmann Viefhaus and Dr. Renè Lehnert for their help, discussions and suggestions. I appreciate Dr. Jorge Antonio Villanueva Garibay for his friendly support, comments and for valuable suggestions. I extend my thanks to Xiarong Yang, Otgontuul Tsetsgee, Natascha Lewanzik, Armin Steinkasserer, Kamalakannan Kailasam and Poonkodi Balasubramaniyan for their friendly support and for providing inspiring working atmosphere.

I am glad to thank Büsnauer Platz friends especially Tushar, Kailash, Venkat, Nitin, Nachi and Ravi for their encouragement, suggestions and help. I wish to thank my school friends, Gobi friends and PSG friends and their families for their friendship, support and for being with me. I am grateful to Sivakumar's dad, Mr. Krishnan, for his invaluable help during my studies.

Many thanks to Ms. Inge Blankenship and Ms. Gisela Hoppe for their friendly support and for helping me with administrative work. I would also like to thank Mr. Walter Ottmüller, Mr. Jochen Graf and Mr. Peter Haller for all mechanical workshop support, Mr. Jürghen Hußke for electronical support and Ms. Gabriele Bräuning and Ms. Annette Hirtler for helping me in the laboratory.

Most importantly, I would like to thank my family. To my grandma (late) P. Vijayalakshmi for her great support and love. I am indebted to you ma for my whole life. To my dad C. Srinivasan, mom S. Lalitha and my sisters S. Bhuvaneswari and S. Anandhi for their encouragement, sacrifices and for everything they have done for me. I would not have reached this position without their well wishes, blessings and prayers. I would like to thank my uncle Mr. G. Elangovan, our family friends Mr. V. Seetharaman and his family and Mr. Ramesh and his family for their great support and well wishers who have helped me and my family.

Last but not least, I take resource of this opportunity to extend my heartfelt gratitude to Almighty God Guruvayurappan who has blessed me with this nice and lovely experience called life.

Contents

Chapter 1

Introduction

Alkyl-bonded stationary phases for reversed-phase high-performance liquid chromatography (RP-HPLC) separations have been developed on the basis of a broad experience and numerous experimental findings by chromatographers. In spite of its popularity as a predominant separation technique, the separation mechanism operative in the chromatographic process has still not been established unambiguously. This is due to the presence of many parameters that control the molecular recognition in the chromatographic process.[1-8] A profound knowledge of the stationary phase morphology is thus a prerequisite to understand solute retention mechanisms, and ultimately column selectivity. Moreover, in order to develop a novel stationary phase for a particular separation, a great number of stationary phases should be synthesized and examined in detail to provide feedback for developing the strategy of the next synthesis. The systematic characterization of stationary phases is thus very important for the development of more powerful stationary phases with desirable separation performance and optimal exploitation of existing materials.

Normal-phase chromatography is defined as an elution procedure, in which the stationary phase is more polar than the mobile phase. This "normal-phase chromatography" term is used in liquid chromatography to emphasize the contrast to reversed-phase chromatography. Typically, stationary phase materials are silica or alumina and the mobile phase consists of a non polar constituent such as hexane modified with a slightly more polar solvent such as chloroform or ethyl acetate. In this method, the more polar compounds are preferentially retained. The reversed-phase mode utilizes the opposite approach for the separation of nonpolar analytes or compounds that have some hydrophobic character. In this case, the mobile phase is significantly more polar than the stationary phase, e.g., a microporous silica-based material with chemically bonded alkyl chains. Here, the mobile phase is composed of a primary polar solvent, usually water, which is modified by a more nonpolar constituent such as methanol, acetonitrile and tetrahydrofuran. The most common types of the RP-HPLC phases are designated by C_n, where n is the number of carbon atoms of bonded linear alkyl hydrocarbon moieties.[8]

In earlier times, RP-HPLC was usually performed using C_{18} or C_8 alkyl modified silica gels. Short chain alkyl phases such as C_2, C_4 and phenyl phases were employed occasionally, while long alkyl chain phases such as C_{30} was used only for specialized applications.[9] Chromatographers believed that they would need only a few stationary phase materials to achieve any separation. Indeed, with the octadecyl-modified system and the ability to adjust different parameters, chromatographers have achieved many successful separations. However, the silica based stationary phases have certain limitations such as poor thermal and chemical stability.[10] Recently, there has been an increasing interest in metal oxides which combines the mechanical strength of silica with the chemical stability of polymer-based stationary phase materials.[11] In fact, stationary phases based on alumina, zirconia and titania were successfully applied in RP-HPLC.[12]

Many approaches have been applied for the characterization of stationary phases. They include systematic analysis of the retention behavior for a certain group of solutes,[13-18] various spectroscopic measurements of the stationary phase materials[4,6,17,19-28] and theoretical calculations[29-32] on the bonded phase structures. In the early stage during the development of the RP-HPLC technique, the stationary phase was considered to play a passive and completely inert role in retention and selectivity. The chain conformations of the stationary phases were primarily studied with the help of chromatographic experiments, which provided essential information about the bonded phase as a function of mobile phase, temperature, solute hydrophobicity and polarity.[14,15,33] These studies have demonstrated that the selectivity differences among the columns are not only due to differences in the starting silica material, but also due to differences in the chain density of the bonded alkyl phase. It turned out that the alkyl chain structure and interfacial properties urge the function and utility of these stationary phase materials. Also it was demonstrated that the conformational order of the alkyl chain moieties plays an important role in determining the efficiency and selectivity of separations. Such studies, however, provide only indirect evidence about the bonded phase morphology.

On the other hand, by using techniques like FT IR,[23,24,34-36] Raman,[25-28] NMR,[6,17,19-22] photo-acoustic,[36] fluorescence spectroscopy,[20] atomic force microscopy[20] and ellipsometry,[20] more direct information about the alkyl chain structure, conformation and dynamics of the stationary phase materials can be attained. A further approach to the study of covalently modified surfaces are computer simulations. Here, suitable simulation models can be set up considering the influence of surface coverage, chain placements and temperature. Simulations can thus provide a further insight into the conformational and dynamic aspects of such stationary phase materials. However, these methods critically depend on the suitability and accuracy of the computational approach employed, i.e., parameterization of the underlying interaction, etc.[29-32]

The application of IR spectroscopy to study alkyl chain conformation was reported by Sander et al.[24] These studies were mainly based on the pioneer work by Snyder et al. on n-alkanes.[37] Since this early work, relatively few efforts have been done for characterization of stationary phase materials using FT IR spectroscopy. More recently, Singh et al.[23] have used FT IR spectroscopy to study the influence of various parameters of alkyl stationary phases like temperature, alkyl chain length and alkyl chain position. Raman spectroscopy provides complementary information in the study of conformational order of alkyl modified surfaces. Dorsey et al.[25,26] and Pemberton et al.[27,28] extensively used Raman spectroscopy to study the influence of temperature, surface coverage, solvents and synthesizing procedure on conformational order of n-alkyl modified systems. Doyle et al.[38] developed an experimental set-up, which permits direct on-column characterization of alkyl stationary phases by Raman spectroscopy under chromatographic conditions.

Solid-state NMR spectroscopy has significantly advanced the development of stationary phase materials. ^{29}Si and ^{13}C CP/MAS NMR techniques provide complementary information on chemically modified surfaces.[4,17,19,20,22] Furthermore, ^{13}C NMR spectroscopy can be used to obtain information about the conformation of surface immobilized alkyl ligands. The dynamics of n-alkyl modified silica gels was studied by using the cross-polarization time constant, T_{CH}, and ^{1}H or ^{13}C spin-lattice relaxation times.[39-41] Pursch et al.[22] used two-dimensional wide-line separation (2D-WISE) NMR to correlate segmental mobility (from ^{1}H line-widths) with chain conformation (from ^{13}C chemical shifts) for C_{22}-alkyl modified silica gels. Another NMR approach which has been used to study the dynamics of modified silica gel is based on NMR line shape changes.[39] The general principle behind the NMR line shape analysis is that rapid molecular motions average dipole-dipole interactions, chemical shift anisotropy, and other anisotropic magnetic interactions resulting in some narrowing of the NMR signals. In addition, Pursch et al.[42] implemented a selective population inversion (SPI) ^{13}C CP/NMR experiment to study the mobility of C_{30} alkyl modified silica gels.

Solid-state ^{2}H NMR spectroscopy on selectively deuterated compounds offers distinct advantages over ^{1}H and ^{13}C NMR methods for probing structure and dynamics in disordered systems.[21,43-45] ^{2}H NMR line shapes and relaxation effects are governed by the quadrupolar interaction, i.e., contributions from chemical shift anisotropy and heteronuclear dipolar interactions as in the case of ^{1}H and ^{13}C NMR spectroscopy, can be neglected. Moreover, the low natural abundance of the deuterium nucleus eliminates background contributions from unlabeled material so that the experimental ^{2}H NMR signal exclusively arises from the deuterons in the selectively labeled molecules. Recently, ^{129}Xe NMR has been used to study the structure, dynamics, and interactions of the stationary phase and the xenon atoms in C_{18} column materials.[46,47]

Although, the results of these aforementioned investigations have afforded valuable insight into the conformational order and mobility of the alkyl chains, a complete molecular picture of stationary phase materials remains elusive. Among the many known variables to affect the retention process, the effect of chromatographic parameters such as substrates, pressure, surface coverage and solvents were least understood. An extensive investigation of stationary phase materials is necessary to address these parameters. Obviously, a single experimental technique is not sufficient to provide a comprehensive and detailed picture about the molecular properties of stationary phase materials.

Therefore, in the present studies both solid-state NMR and FT IR techniques are employed to examine the conformational order and mobility of alkyl chain segments chemically attached to solid surfaces. The desired information is obtained by the analysis of various conformational-sensitive IR bands, including CH_2 and CD_2 stretching, CH_2 wagging, and CD_2 rocking modes.[37,48] A qualitative statement about the conformational order is based on the analysis of symmetric and anti-symmetric CH_2 and CD_2 stretching bands. The frequency shift of the band maxima in the symmetric/anti-symmetric regions provides information about the changes in the conformational order that may result from changes in surface coverage, sample temperature, etc. The analysis of the CH_2 wagging bands regions between 1330 and 1400 cm^{-1}

provides the relative amounts (i.e., integral numbers over the whole chain) of the kink/gauche-trans-gauche, double-gauche and end-gauche conformers in the tethered alkyl chains. From the analysis of the CD_2 stretching and CD_2 rocking bands, information about the conformational order at a specific deuterated methylene segment is available. Here, CD_2 rocking band data are used to determine the amount of gauche conformers at the deuterated carbon positions C-4, C-6, and C-12 in alkyl chains of different lengths. The silane functionality and degree of cross-linking of silane ligands on the silica surface are evaluated by ^{29}Si CP/MAS NMR while the structural order and mobility of the alkyl chains are investigated by ^{13}C CP/MAS NMR spectroscopy.

In the present work, the influence of substrate on alkyl chain conformational order of C_{30} and C_{18} stationary phase materials are examined. The impact of the external pressure during sample preparation on the alkyl chain conformations is evaluated for non-deuterated and selectively deuterated alkyl modified silica gels in the dry state by variable temperature FT IR spectroscopy. In order to examine the impact of surface coverage on the conformational order of stationary phase materials, C_{18}-alkyl modified silica gels with variable surface coverages are synthesized and are characterized by the aforementioned spectroscopic techniques. The obtained molecular behavior of tethered alkyl chains is further compared with the chromatographic performance of these phases, in terms of the shape selectivity data. Likewise, the influence of the synthetic routes (solution and surface polymerization techniques) on the molecular properties of tethered alkyl chains is examined on two representative samples, namely C_{18} and C_{30}-alkyl modified silica gels. In the final part, the influence of solvents on the conformational order C_{18}-alkyl modified silica gels is investigated to probe the interactions of the mobile phase with the stationary phase materials and their impact on the conformational order of alkyl modified stationary phase materials.

Chapter 2

Systems

Reversed-phase high-performance liquid chromatography (RP-HPLC) is the most popular mode of high-performance liquid chromatography (HPLC). The major advantage of RP-HPLC over other HPLC techniques such as ion-exchange (IE-HPLC) or normal-phase chromatography (NPC) is due to its versatility, familiarity to chromatographers, the number and variety of commercial phases available, and the large number of applications. Reversed-phase chromatography can achieve a larger range of separations than all other modes, because it enables users to control the mobile phase by changing solvent type, solvent composition, and pH by adding modifiers such as surfactants, chiral reagents, competing bases, and ion pair reagents or by adjusting experimental conditions such as flow-rate and temperature.[49,50]

Silica was the most common material used in the early development of column liquid chromatography. However, silica is a polar material, which contains hydroxyl groups (silanols) which are acidic as well as strongly hydrogen bonding in character. These properties make it inappropriate as a stationary phase for many typical organic molecules that are predominantly hydrophobic. Moreover, the silanols interact strongly with basic compounds leading to poor chromatographic results. The modification of the surface is thus crucial to provide a more nonpolar (hydrophobic) material which does not lack of these undesirable effects of silica and to have a medium more suitable for the separation of a large variety of organic compounds. It is beneficial to keep silica as the primary material in the column because it has the ability to be modified either chemically or physically by adsorption. Physical adsorption has been used to modify silica surfaces for chromatographic purposes. However, modification by physical adsorption is limited because of the nature of modern

HPLC. The use of high pressure creates shear forces at the interface between the stationary and mobile phases and is high enough to remove even insoluble liquids from the surface of the solid support. The stationary phase is then forced out of the column as an insoluble droplet. Removal of the stationary phase from a chromatography column is usually referred as "column bleed". Therefore, chemical modification is solely an approach to modify the silica surface in order to create a stationary phase, which is compatible with the types of solutes to be separated. [1,8,51,52]

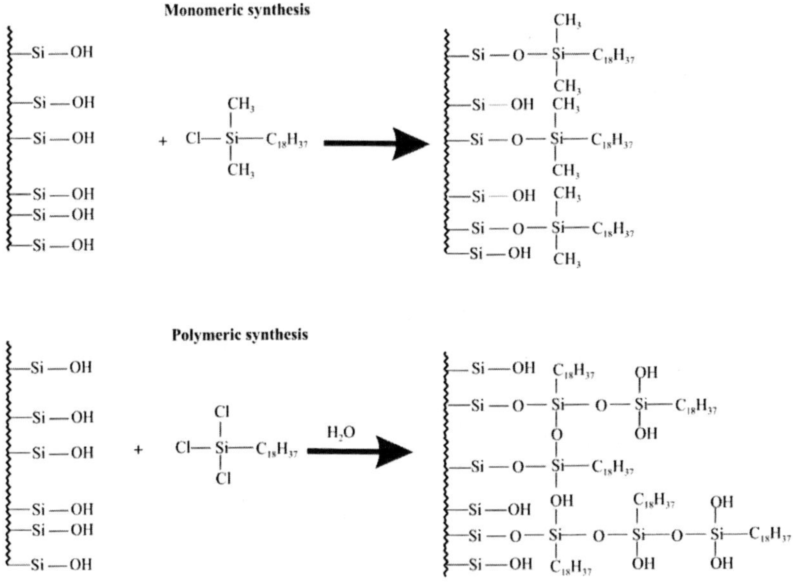

Figure 2.1: Conventional synthetic schemes for monomeric and polymeric alkyl modified phases.

Alkyl chains attached to silica gels are one among the popular stationary phases synthesized by covalent attachment of alkyl chlorosilanes to a solid support using mono-, di- or tri-chlorosilanes. Monomeric stationary phase materials can be obtained by the attachment of alkyl chains with silica through single bonds. When di- or tri-functional silanes are used, two or more bonds per ligands can be attained (Figure 2.1). The procedures used for the polymeric phase synthesis have been described as "solution polymerized" and "surface polymerized" to distinguish the introduction of water for initiating the polymerization (Figure 2.2). For solution polymerization, water is added to a slurry of silica containing di or trichlorosilane. Polymerization occurs in the solution with subsequent deposition onto the silica. Deposition of the silane polymer on the silica surface would result in a surface with some heterogeneity. For surface polymerization, water is added to dry silica either through exposure to humid air or by direct addition prior to silanization. Here, a monolayer coverage of the water molecules on the silica surface is achieved. Later, wet silica is introduced into a solution containing the silane. The "surface polymerized" procedure is a self-assembled monolayer approach, where higher surface coverage of alkyl chains on the surface and a more regular bonded surface can be reached.[19,53] However, because of steric hindrance, only a fraction of the silanol groups (<50%) participate in the reaction.

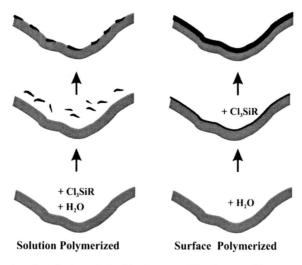

Solution Polymerized   **Surface Polymerized**

Figure 2.2: Reaction scheme for the synthesis of the chromatographic column materials.

Alkyl modified stationary phases offer a wide range of chromatographic performance which result from differences in substrate and stationary phase properties, such as surface area, pore size and pore size distribution, particle size and particle size distribution, porosity and chemical composition,[2,3,51,54] bonding chemistries (monomeric, solution polymerized, surface polymerized),[17,55] surface coverage,[17,55] mobile phase,[25,27] temperature,[23,26] solute hydrophobicity and polarity.[14,15,33] For example, highly ordered, high-density stationary phases have shown to possess enhanced shape selectivity for planar polycyclic aromatic hydrocarbons (PAHs), while less ordered stationary phase materials are characterized by reduced geometric selectivity.[56,57]

Silica based stationary phases have been traditionally used as a substrate material for RP-HPLC due to their high mechanical strength, narrow particle size distribution, high specific surface area, variance of pore size and well-documented chemistry of surface modification (see Figure 2.3).[52] However, the silica based stationary phase materials have certain limitations such as remaining adsorptivity towards basic (amines) compounds due to silanol interactions, leading to asymmetric peaks. They also lack of pH stability because the silica backbone Si-O-Si hydrolyses at pH > 8, and the siloxane bond is unstable at pH < 2. Use of column temperatures above 60 °C can also result in hydrolyses of the stationary phase from silane-modified silicas.[10] In spite of the mentioned limitations, until now silica has remained as the one of the major substrate for RP-HPLC stationary phases and the majority of the presently applied RP-HPLC separations are performed on silica-based materials.

Many attempts have been undertaken to overcome these drawbacks of silica-based materials, for instance, by covalently bonding alkyl chains, modification with a polymeric reagent or coating with an inorganic oxide. Another solution to overcome the drawbacks of silica is substitution by an oxide with greater pH stability than silica. Titania, zirconia and alumina could provide an alternative to silica as chromatographic packing materials, and give rise to high mechanical stability, separation efficiency, and chemical inertness at elevated temperatures and over a wide pH range of 1 to 14.[54] Unger et al.[52,58] have compared the chromatographic performance of zirconia, silica, alumina and titania in the normal phase HPLC mode and showed that zirconia is superior with respect to the separation of basic compounds. Nawrocki et al.[54] studied the physical and chemical properties of microporous zirconia that exhibited excellent properties as stationary phase material. In principle, zirconia based chromatographic phases thus have the potential to overcome the drawbacks of silica.[59]

Combined solid-state NMR and chromatographic studies have provided a deeper insight into the molecular shape selectivity of different n-alkyl modified stationary phase materials.[17,42] A considerable effort has been devoted to alkyl modified silica gels, especially C_{18}-alkyl modified silica gels since it is hydrophobic enough to react with almost all organic compounds to some extent. For such alkyl modified stationary phase materials, the conformational order of the alkyl chain moieties is intimately linked with the selectivity during chromatographic separations.[16,53,60] In earlier studies by Sander and Wise,[16] C_{18} stationary phases were synthesized by varying the monomeric and polymeric surface modification procedures to evaluate the relationships between bonding chemistry and chromatographic properties. Distinct differences in shape selectivity among the columns were observed and it was concluded that selectivity differences result from differences in the alkyl chain conformational order as a consequence of alkyl chain organization and surface coverage. Pursch et al.[17] prepared C_{18} phases by different synthetic pathways which were then examined by solid-state NMR spectroscopy and liquid chromatography. It was concluded that the C_{18} phases prepared by the surface polymerization technique exhibit a more regular surface coverage and consequently enhanced shape recognition than phases prepared via solution polymerization. In contrast, very few studies of long chain (C_{30}) SAMs have been reported so far.[61-63] Silica based C_{30} phases designed for liquid chromatography have received increased attention after the successful application to the separation of carotenoids, vitamin A isomers, fullerenes, tocopherols, and tocotrienols.[64-66]

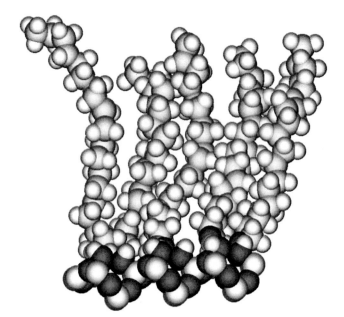

Figure 2.3: Pictorial view of C_{18}-alkyl modified silica gels.

One of the least understood aspects of retention mechanism are intermolecular interactions between each chromatographic component such as bonded alkyl chain, solvent and solute over the course of separation. Such interactions include solvent-solute, solute-stationary phase, solvent-stationary phase and also the intermolecular interactions of the alkyl chains with in the stationary phases. The effect of solvent on the alkyl chain conformational order has been examined in numerous studies.[24,25,27,45,67-69] Two models of solute retention process have been proposed, where the importance of stationary phase structure varies significantly. In the solvophobic model (hydrophobic model or adsorption model), the stationary phase plays a passive role providing a sorption site for the molecules to interact. In this model, retention is treated simply as an adsorption-desorption process.[70-73] In the partitioning model, the solute is embedded in a cavity formed between the chains of the stationary phases. The stationary phase in this model plays an active role in retention with solute partitioning inducing changes in conformational order of the alkyl chains of the stationary phases.[74-77] Although these models specifically address the interaction between solute and stationary phase, solvent-stationary phase interactions should be examined. It is important to note that these are theoretical models of interactions and neither model may be correct or completely describe a particular retention process.

Elemental analysis is one of the basic methods for evaluating stationary phases.[78,79] The quantity of carbon, nitrogen, and hydrogen can be measured directly by burning the sample in oxygen. It was widely utilized to provide an indication of surface coverage. In combination with specific surface area of the substrate, carbon analyses can be used to calculate the surface coverage in $\mu mol/m^2$ or $groups/nm^2$. However, elemental analysis does not provide information about the homogeneity of the stationary phase and can be used only as a complement to other more sophisticated analytical characterization methods.

Although the results of these investigations have provided valuable insight into the retention process, a clear understanding of the complex architecture of alkyl modified stationary phases remains a challenge. Describing the alkyl chain conformational order on the molecular level is thus crucial for the understanding of such materials in their respective chromatographic applications.

In the present work, considerable efforts are made in order to get a better insight into the morphology of the alkyl chain structure. The influences of substrates, pressure, surface coverage, synthesis procedure and solvents on the alkyl modified systems are examined here.

2.1   Influence of Substrates

FT IR spectroscopy is used to examine the conformational order of alkyl chains in four different C_{30} self-assembled monolayers (SAMs) which differ by the respective solid substrate. Thus, C_{30} SAMs were prepared by reacting $C_{30}H_{61}SiCl_3$ with the humidified surfaces of zirconia, titania, and two different silica gels through the "surface polymerization technique" or self-assembled monolayers approach. In addition, the conformational order of alkyl chains of commercially available carbon clad zirconia based supports and synthesized C_{18}-alkyl modified silica based supports (see Figure 2.4) are probed in the dry state using variable temperature FT IR and solid-state ^{13}C NMR spectroscopy.

2.2   Influence of Pressure

The conformational order of specific chain positions as a function of the alkyl chain lengths are probed using selectively deuterated C_9H_{19}-, $C_{18}H_{37}$-, $C_{22}H_{45}$- modified silica gels (see Figure 2.4). In addition, non-deuterated alkyl modified silica gels are also included. Particular emphasis is given to the impact of the external pressure during sample preparation on the alkyl chain conformations, of which there is no information available so far. The desired information is obtained by the analysis of various conformational-sensitive IR bands, including CH_2 wagging, CD_2 stretching and CD_2 rocking bands.

2.3   Influence of Surface Coverage

C_{18}-alkyl modified silica gels with different surface coverage varying from 2 to 8.2 $\mu mol/m^2$ are prepared by different synthetic pathways and are examined by FT IR, solid-state NMR spectroscopy. A detailed variable temperature FT IR study is carried out for the first time by dealing with the influence of surface coverage on the conformational order of C_{18}-alkyl modified silica gels. The silane functionality and the degree of cross-linking of silane ligands on the silica surface are evaluated by ^{29}Si CP/MAS NMR measurements while the structural order and mobility of the alkyl chains are investigated by ^{13}C CP/MAS NMR spectroscopy. The derived results from the FT IR and NMR are discussed in the context of chromatographic shape selectivity differences in order to get a better insight of the morphology and alkyl chain structure of the stationary phase materials.

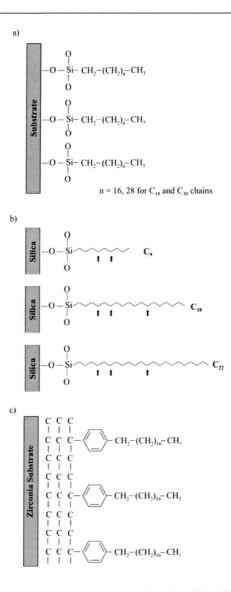

Figure 2.4: Molecular structures of the investigated systems: a) C_{18} and C_{30}-alkyl modified silica gels, b) selectively deuterated n-nonyl and n-octadecyl and n-docosyl modified silica gels, where arrows indicate the deuterated positions, c) commercial column material (DiamondBond®-C_{18}).

2.4   Influence of Synthetic Routes

Modified silica gels with n-alkyl chains (n = 18, 30) are prepared by solution and surface polymerization techniques (see Figure 2.4) and are examined in the dry state by variable temperature FT IR and solid-state NMR spectroscopy. The silane functionality and degree of cross-linking of silane ligands on the silica surface are determined by ^{29}Si CP /MAS NMR measurements. The structural order and mobility of the alkyl chains over a broad temperature range are investigated by ^{13}C CP/MAS NMR measurements. Variable temperature FT IR studies are performed from which conformational order and flexibility of the alkyl chains in C_{18} and C_{30} phases are monitored through conformational sensitive CH_2 symmetric, anti-symmetric stretching and wagging modes.

2.5   Impact of Solvents

In addition, the influences of solvent and sample temperature are studied for C_{18}-alkyl modified silica gels. Here, various solvents such as acetone, chloroform, cyclohexane, dichloromethane, perdeuterated solvents (chloroform, n-hexane, DMF, cyclohexane) were added to the dry C_{18}-alkyl modified silica gels. These samples are then examined by variable temperature FT IR spectroscopy in order to study the conformational order of alkyl modified silica gels under chromatographic conditions.

Chapter 3

Spectroscopic Techniques

During the course of these studies, solid-state NMR and FT IR spectroscopic techniques have been employed since both can be used to get complementary information of the tethered alkyl chains attached to the different metal oxides. FT IR and NMR are distinguished by their experimental time-scales. As FT IR spectroscopy probes on a time scale of less than 10^{-10} s, it is perfectly suited as a snapshot for studying the population of different conformers for a given area of irradiation, while NMR spectroscopy provides an insight into the motional processes which accompany the chain disordering for instance, in stationary phase materials. A major advantage of FT IR spectroscopy is given by its high sensitivity and relatively easy access to the desired information about the conformational order. However, often FT IR spectroscopy is not applicable for the study of the alkyl conformational order, since the vibrational bands are obscured by other strong FT IR bands of the materials under investigation. In principle, NMR spectroscopy is superior where the information about both the chain dynamics and the chain ordering is accessible. On a qualitative basis, the chain ordering is obtained via the analysis of ^{13}C NMR spectra where "trans" and "gauche" NMR signals can be clearly distinguished.[4,20] However, quantitative data about these parameters are only available after performing a detailed data analysis of the NMR line shapes or relaxation data.[39] Thus the application of FT IR and NMR techniques can be used to get complementary information of systems bearing n-alkyl chains.

3.1   Fourier Transform Infrared Spectroscopy

FT IR is a powerful and versatile analytical method, which has been used as a conventional technique for material analysis for over seven decades. An infrared spectrum represents a fingerprint of a sample with absorption peaks that correspond to the frequencies of vibrations of the bonds between the atoms in the systems. Since each different material is a unique combination of atoms, no two compounds produce the same exact infrared spectrum. Therefore, infrared spectroscopy can be used for identification of materials.

In the past, it was demonstrated that the observed absorption frequencies, intensities and band shapes critically depend on the molecular conformation, configuration and chain packing.[48] It is well known that even highly crystalline polyethylene exhibits certain absorption bands, which cannot be assigned to the vibrational modes of infinitely long and fully extended all-trans chains. These bands are associated with vibrations of non-planar chains (gauche conformers). The randomness in the chain conformation tends to promote the localization of some vibrational modes, which results in the spectrum as well-defined bands associated with specific conformations.[37]

Snyder calculated and tabulated the IR frequencies for an extensive number of vibrations in alkane systems, including some that are specific to localized bent structures containing gauche bonds. The normal mode calculations detail the assignment of various C-H stretching, scissoring, rocking, and wagging modes. The intensities and positions of these bands provide quantitative information about various defect structures within the chains that contribute to the overall disorder of the system.[37]

For instance, the position of the CH_2 symmetric and anti-symmetric stretching band maxima (from 2853 to 2846 cm^{-1} and 2926 to 2912 cm^{-1}, respectively) and CD_2 symmetric and anti-symmetric stretching (from 2090 to 2100 cm^{-1} and 2195 to 2200 cm^{-1}, respectively) were used to study the conformational order of n-alkanes. In addition, the wagging band progressions (1100 - 1350 cm^{-1}) could provide useful insight into the chain conformation on a qualitative basis. Moreover, the wagging and rocking modes of methylene and CD_2 groups give rise to conformation sensitive bands in the regions from 1330 to 1370 cm^{-1} and from 610 to 670 cm^{-1}, respectively, yielding quantitative information about the conformers present in n-alkanes. In recent years, the aforementioned IR vibrational modes have been used to study the conformational order of alkyl modified stationary phases. Sander et al.[24] reported the first application of FT IR in the study of the conformational structure of C_1 to C_{22}-alkyl modified silica surfaces though semiquantitative assessment of C-H stretching and wagging mode bands. Since this early work, relatively few efforts have utilized FT IR to characterize conformational details of chromatographic stationary phases. Jinno et al.[34,35] reported differences in alkyl chain order and rigidity for monomeric and polymeric C_{18} stationary phases, as assessed with diffuse reflectance FT IR, and Lochmüller et al.[36] demonstrated the use of FT IR photoacoustic spectroscopy to characterize octadecyl-modified surfaces. Recently, Singh et al.[23] described a comprehensive FT IR investigation of alkyl chain conformational order in alkyl modified silica gels. Here, the influence of alkyl chain length, alkyl chain position and temperature were examined.

For the present studies, several vibrational modes are used to investigate the conformational order of the alkyl modified stationary phase materials. The information related to each investigated vibrational mode is provided in Table 3.1.

Table 3.1: Conformational-sensitive IR modes used for the present investigations.

Vibrational mode	Wavenumber (cm^{-1})	Comments
CH_2 stretching:		Shifts to higher frequency indicate more disorder.
Symmetric	2853-2846	Gauche amount (qualitative)
Anti-symmetric	2926-2912	
CD_2 stretching:		Shifts to higher frequency indicate more disorder.
Symmetric	2115-2070	Gauche amount (qualitative)
Anti-symmetric	2200-2165	
CH_2 wagging	1370-1330	Integrated peak intensity is correlated with conformation. gtg`& gtg, gg, tg (at chain end) Gauche amount (quantitative)
CD_2 rocking	670-610	gtgt, gtgt, ttgt, tt, Gauche amount (quantitative)

3.1.1 Methylene Stretching

The position of the CH_2 symmetric and anti-symmetric stretching band maxima (from 2853 to 2846 cm^{-1} and 2926 to 2912 cm^{-1}, respectively) provides qualitative information about the changes in the conformational order as a function of external parameters, such as surface coverage or sample temperature. For completely disordered "spaghetti like" structures, the frequency of the CH_2 anti-symmetric stretching band is close to that of a liquid alkane ($\tilde{\nu}_a = 2928$ cm^{-1}). For well-ordered systems, the frequency is shifted to lower wavenumbers and is close to that of crystalline alkanes ($\tilde{\nu}_a = 2920$ cm^{-1}).[80] A similar shift can be registered for the CH_2 symmetric stretching band as well.

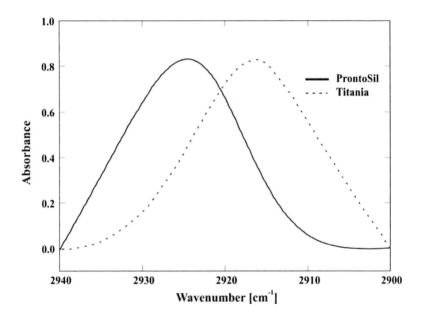

Figure 3.1: Experimental CH_2 stretching bands for C_{30} SAMs on silica (ProntoSil) and titania at 353 K.

Figure 3.1 illustrates the FT IR spectrum for the C_{30} SAMs in the anti-symmetric stretching band region between 2940 and 2900 cm^{-1}. The CH_2 anti-symmetric stretching band maximum varies by 10 cm^{-1} between the C_{30} SAMs on titania and the C_{30} SAMs on silica, which reflect significant differences in conformational order for these materials. Another example shows the exploitation of CH_2 stretching bands in order to study the influence of temperature on the conformational order of alkyl modified systems (see Figure 3.2). In addition, the CH_2

stretching bandwidth can be utilized to study the alkyl chain flexibility of the stationary phase materials. Here, the CH_2 bands are narrow at lower temperatures, reflecting a low alkyl chain flexibility and high conformational order.[81] At higher temperatures, an increase of the CH_2 stretching bandwidth is observed which can be attributed to an enhanced alkyl chain flexibility owing to a decrease of conformational order.

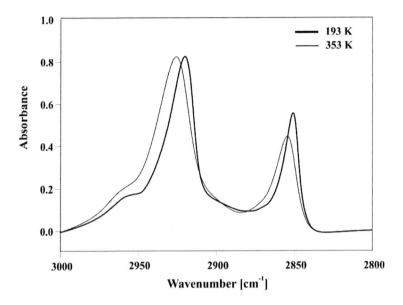

Figure 3.2: Temperature dependence of CH_2 stretching modes of C_{30} SAMs on silica (ProntoSil).

3.1.2 Methylene Wagging

Quantitative information of conformational order of alkyl chain containing systems can be attained through an analysis of the CH_2 wagging band regions. The basis of this conformational study is given by the pioneering work of Snyder et al.[37] who calculated and observed the vibrational frequencies for CH_2 wagging band modes in protonated liquid n-alkanes containing specific nonplanar conformers. In Figure 3.3, a representative experimental FT IR spectrum covering the CH_2 wagging band region between 1300 and 1400 cm^{-1}, recorded at 253 K, is shown along with the theoretical curves from the curve fitting analysis. The intensities of these bands can be used to obtain quantitative information on various "defect" structures within the chains that contribute to the overall disorder of the system. The conformation-dependent wagging modes of interest, appear near to 1368 cm^{-1}, 1353 cm^{-1} and 1341 cm^{-1} and arise from kink/gtg, double-gauche and end-gauche sequences, respectively, as shown in Figure 3.4. The kink/gtg conformers are taken together since it is not possible to distinguish spectroscopically from the CH$_2$ wagging bands. The kink/gtg sequence provides some disruption in the order of the interior segments of the chain, while the double-gauche sequence creates a severe destruction from the alignment of chains in the all-trans conformation. In contrast, the end-gauche sequence provides the least perturbation to chain order, since it occurs only at the terminal methyl end groups of the chains. For the analysis of such wagging bands, the band intensities in the experimental spectrum are normalized with respect to conformational insensitive methyl group umbrella deformation mode at 1378 cm^{-1}. It is noteworthy to mention that the frequencies of the wagging bands due to the above-mentioned conformers appear within a few wavenumbers from each other. Therefore, curve-fitting analyses are employed to extract the intensities of the respective bands in the experimental spectrum.

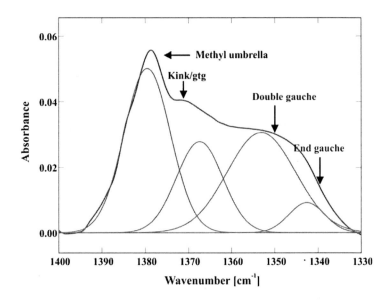

Figure 3.3: Experimental FT IR wagging mode of C$_{18}$-alkyl modified silica gel at 253 K along with the theoretical curves from the curve fitting analysis.

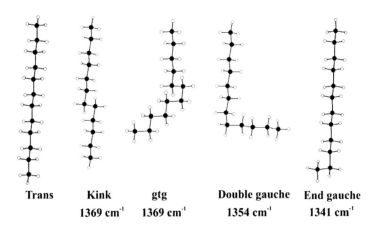

Trans	**Kink**	**gtg**	**Double gauche**	**End gauche**
	1369 cm⁻¹	1369 cm⁻¹	1354 cm⁻¹	1341 cm⁻¹

Figure 3.4: Alkyl chain conformations. The non-planar conformations give rise to localized mode vibrations, which are observed in FT IR spectra (wagging band region).

The amount of specific gauche sequences was calculated based on reference measurements on alkanes and a theoretical approach using the rotational isomeric state (RIS) model given by Flory.[82] The RIS model predicts the numbers of the three conformations expected at a given temperature for n-alkanes of a given chain length. Calculations were performed for C_{14} to C_{18} in a temperature range from 273 to 373 K by Senak et al.[83] A cubic polynomial function was fitted to each of the calculated percentages of double-gauche (represented as gg), kink (gtg`) and gtg (designated as k), and end-gauche (eg) conformers as a function of temperature. The calculated percentages, i.e., P_{gg}, P_{eg} and P_k, which are independent of chain length in accordance with the RIS model, are given as follows:

$$P_{gg}(\%) = -7.364 + 8.859 \times 10^{-2}\,T - 1.388 \times 10^{-4}\,T^2 + 8.741 \times 10^{-8}\,T^3 \tag{3.1}$$

$$P_k(\%) = -9.092 + 12.49 \times 10^{-2}\,T - 2.732 \times 10^{-4}\,T^2 + 2.176 \times 10^{-7}\,T^3 \tag{3.2}$$

$$P_{eg}(\%) = -4.522 + 1.620 \times 10^{-1}\,T - 3.486 \times 10^{-4}\,T^2 + 2.855 \times 10^{-7}\,T^3 \tag{3.3}$$

In these three equations, T is the temperature in Kelvin and the calculation is valid (for interpolative purposes) from 273 to 373 K. P_{gg} and P_k include distinct counting of conformers of the type (gg, g`g) or (g`tg, gtg`, g`tg`and gtg), respectively, while end-gauche calculations refer to a single state only. The percentage probabilities of these conformers can be converted to their absolute numbers per chain for a particular n-alkane molecule using the following equations:

$$\text{No. of double } gauche \text{ conformers} = P_{gg}\left\{\frac{N_{c-c}-3}{100}\right\} \tag{3.4}$$

$$\text{No. of } (kink + gtg) \text{ conformers} = P_{k}\left\{\frac{N_{c-c}-4}{100}\right\} \tag{3.5}$$

$$\text{No. of end } gauche \text{ conformers} = \left\{\frac{4P_{eg}/100}{1+P_{eg}/100}\right\} \tag{3.6}$$

where Nc–c is the number of C–C bonds in the n-alkane. The number of end-gauche conformers is chain length independent, since a maximum of two are possible regardless of the chain length.

According to this model, the various integrated wagging bands intensities are determined by curve-fitting of the measured spectrum after baseline corrections. The resultant bands are then normalized to the intensity of the methyl umbrella mode. The amount of specific gauche sequences in the n-alkyl modified system is then calculated using the reference plots of n-alkanes from Neumann-Singh et al.[84] which is based on the RIS model (using Equations 3.1 to 3.6).

The number of conformers per chain was calculated based on the fact that the surface linked alkyl chains contain only one methyl group per chain, in contrast to liquid n-alkanes which contain two methyl groups per chain. The total number of gauche conformers per chain was obtained by taking into account that two gauche bonds are necessary to comprise one kink or double-gauche sequence and only one gauche bond for the formation of an end-gauche sequence.

3.1.3 CD$_2$ Stretching

Similar to methylene stretching bands, qualitative information on the conformational order for a specifically deuterated methylene segment is obtained via the position and widths of the CD$_2$ symmetric (2115 to 2070 cm^{-1}) and anti-symmetric stretching bands (2200 to 2165 cm^{-1}). The main advantage is that the obtained information through CD$_2$ stretching bands is more specific rather than information about the conformational order of whole alkyl chains with the help of CH_2 stretching bands.

3.1.4 CD₂ Rocking

The amount of gauche conformers at a specific site in the alkyl chain containing systems is accessible via the analysis of CD_2 rocking bands of selectively deuterated samples. This method is based on the work by Snyder et al.[48] in which they utilized CD_2 rocking modes to probe trans-gauche isomerization at a specific location in a partially deuterated polyethylene chains. The approach relies on the fact that CD_2 rocking modes in a CH_2-CD_2-CH_2 skeleton appear at 622 cm^{-1} if the local geometry is trans-trans (tt), but shifts to about 650 cm^{-1} upon formation of trans-gauche conformers (tg) with secondary shifts resulting from conformational effects farther along the chain. When the conformations are gauche-gauche, the CD_2 rocking frequencies appear at about 680 cm^{-1}. However, this 680 cm^{-1} band is not easily detectable due to the overlap of other stronger bands and thus is rarely used for conformational analysis.

In the case of liquid n-heptane-4-d_2, conformational order is manifested by the appearance of a pair of tg bands (near 649 and 644 cm^{-1}), instead of the expected single band.[48] This could be due to the next nearest CH_2 neighbours of the CD_2 group that are involved to a sufficient extent to affect the CD_2 rocking frequency, and thus provide conformational information farther away from the tagged methylene. As a result, these bands are used as conformational markers of sequences gtgt or g`tgt which appear at 644 cm^{-1}, and the ttgt sequence which appears at 649 cm^{-1}.

A representative spectrum covering the CD_2 rocking band region for C_{18}-alkyl modified silica gels deuterated at position C-4 (C_{18}Si-4) at 193 K is shown in Figure 3.5. In contrast to liquid n-heptane, a single band is observed at 651 cm^{-1}, which can be used as a direct measure of gauche conformers. The integral intensities of these two bands after iteration can be used to calculate the percentage of gauche conformers at the specific deuterated position in the chain segment using the following equation:

$$\%gauche = \frac{I[651\ cm^{-1}]}{2 \times I[622\ cm^{-1}] + I[651 cm^{-1}]} \qquad (3.7)$$

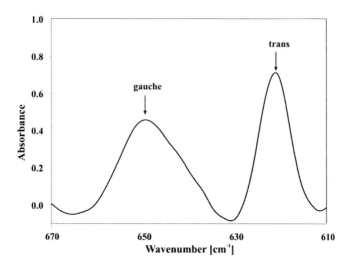

Figure 3.5: Representative FT IR spectra (CD_2 rocking band region) of n-octadecyl modified silica gels selectively deuterated at the C-6 position at 273 K.

3.2   Solid-State NMR Spectroscopy

NMR spectroscopy offers a comprehensive approach to the study of alkyl modified stationary phase materials. Solid-state NMR measurements of chromatographic materials yield extensive information about the morphology of dry bonded phases, whereas suspended-state NMR approaches provide details of the motion and conformation of bonded species in the presence of a solvent. Solid-state NMR has been proven to be a powerful technique for characterizing the stationary phase materials, in particular for studying the alkyl chain mobility and conformational order of the stationary phases. In the early 1980s, Maciel and Sindorf pioneered the application of solid-state NMR techniques in chromatographic stationary phase materials by examining silica surfaces with ^{29}Si CP/MAS NMR spectroscopy.[85] Later, they explored the structure and mobility of tethered monofunctional C_8 and C_{18} ligands with ^{13}C CP/MAS NMR spectroscopy.[40] The combination of cross polarization (CP) with magic angle spinning allows acquisition of high-resolution NMR spectra of low-abundance heteronuclei like ^{13}C and ^{29}Si in reasonable measuring times. Quantitative comparisons of ^{29}Si CP/MAS NMR signals are not possible since the efficiency of magnetization transfer by CP is highly variable. However, ^{29}Si and ^{13}C NMR techniques provide detailed information about the different groups present in substrate and chemically modified surfaces. ^{2}H NMR spectroscopy can be applied to the study of alkyl chain motion of different segments of the alkyl chain through selective deuteration.[21,43-45]

3.2.1 ^{29}Si CP/MAS NMR Spectroscopy

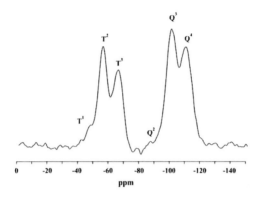

Figure 3.6: ^{29}Si CP/MAS NMR spectrum of C$_{18}$-alkyl modified silica gels with surface coverage of 8.2 μmol/m^2.

^{29}Si CP/MAS NMR can be used to determine and confirm the functionality of the silanes in the alkyl modified substrates. A representative ^{29}Si CP/MAS NMR spectrum of C$_{18}$-alkyl modified silica gel of surface coverage (8.2 μmol/m^2) is shown in Figure 3.6. This spectrum reveals the chemistry of surface modification reaction employed in the synthesis of alkyl bonded phases. Because of the chemical shift dispersion of about 130 ppm for silane and silica gel signals, structural elements can be assigned quite easily. The presence of various types of silicon atoms as a result of the silylation step can be ascertained from this figure. The different structural elements are illustrated in Figure 3.7. The signals from the native silica are Q^2, Q^3, Q^4 and are located in the range from -91 to -110 ppm, in which superscripts indicate the number of Si-O-Si bonds. In contrast to FT IR spectroscopy, where isolated and geminal silanols absorb at nearly the same wavenumber, Q^3 units (silanol groups) and Q^2 (geminal silanols groups) can be distinguished by their corresponding chemical shifts at -91 and -100 ppm, respectively. The ^{29}Si NMR signals of the trifunctional silanes (T) are located in the range from -45 to -70 ppm which refer to tri-functional groups without cross-linking (T^1) at about –48 ppm, with partial (T^2) at about –55 ppm, and with complete cross-linking (T^3) in the range from –66 to –68 ppm.

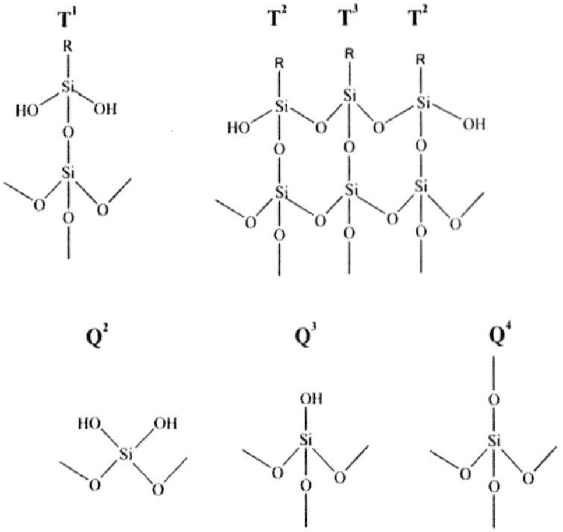

Figure 3.7: Structural elements of the different silicon species present in the C_{18}-alkyl modified silica gels.

3.2.2 ^{13}C CP/MAS NMR Spectroscopy

Carbon nuclei offer a wider chemical shift range (220 ppm) than protons (12 ppm) and therefore provide a sensitive indicator of structural changes and the conformational order of immobilized ligands. Figure 3.8 shows the ^{13}C CP/MAS NMR spectra of C_{18}-alkyl modified phases with different surface coverages.

In solid-state ^{13}C NMR study, conformational order is typically inferred from the relative intensity of the main methylene carbon resonance assigned to "crystalline like" trans and "solution-like" mixture of trans and gauche conformations.[4,20] The ^{13}C CP/MAS NMR spectra show substantial differences for the conformational sensitive ^{13}C resonance of the inner methylene units. The signal position for 2 μmol/m^2 sample is observed at 30.6 ppm, which indicates a high proportion of gauche conformers, while in the case of 8.2 μmol/m^2, the ^{13}C resonance appears at 33.5 ppm, reflecting the presence of chains in a predominantly trans conformational state.

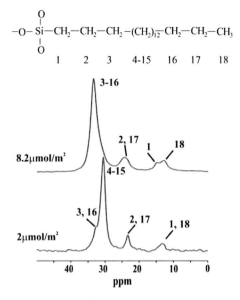

Figure 3.8: ^{13}C CP/MAS NMR spectra of C_{18}-alkyl modified silica gels.

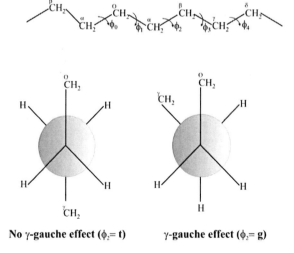

Figure 3.9: Newman projections illustrating the γ-gauche effect.

The lowfield shift in the ^{13}C CP/MAS NMR spectra of the present phases can be explained through the well-known γ-gauche effect (see Figure 3.9). Thus, the distance between the

27

observed carbon ^OC and its γ-substituent $^\gamma C$ for an alkyl chain segment $^OCH_2CH_2$-$CH_2^\gamma CH_2$ depends on the conformation of the centre bond. The distance is reduced from 4 to 3 Å when the conformation changes from the trans to the gauche state along with a change of the electron shielding at position ^OC, causing an upfield shift in the respective ^{13}C NMR resonance.[86,87] Changes in alkyl chain conformation that result from changes in surface coverage, alkyl chain length, and temperature thus can be evaluated from the conformational sensitive ^{13}C resonance of the inner methylene units.

Chapter 4

Experimental Section

4.1 Materials and Synthesis

4.1.1 C_{30} Self-Assembled Monolayers

The self-assembled monolayer materials were synthesized by Dr. Lane Sander, NIST, USA. The essential parameters and properties of the C_{30} self-assembled monolayers are summarized in Table 4.1.

Table 4.1: Sample properties of the C_{30} SAMs for studying the effect of substrates.

Support	Mean pore size (Å)	Specific surface area (m^2/g)	% C	Surface coverage $(\mu mol/m^2)$
Silica (LiChrospher)	300	77	8.77	3.57
Silica (ProntoSil)	260	109	16.77	5.22
Titania (Sachtopore)	286	17	3.58	6.10
Zirconia (Zirchrom phase)	290	31	6.51	6.44

4.1.2 DiamondBond®-C$_{18}$ Column Materials

The DiamondBond®-C$_{18}$ column materials are commercial stationary phase materials developed by Cabot Corporation and Zirchrom Separations Inc. The level of surface coverage with C$_{18}$ groups can be varied by changes in the stoichiometry of the surface modification reaction. In the present work, two different DiamondBond®-C$_{18}$ column materials are studied, i.e., (i) low density DiamondBond®-C$_{18}$ (hereafter denoted as LDZr-C$_{18}$) with a surface coverage of 2.5 μmol/m^2, and (ii) high density DiamondBond®-C$_{18}$ (hereafter denoted as HDZr-C$_{18}$) with a surface coverage of 4.0 μmol/m^2.

4.1.3 Selectively Deuterated n-Alkyl Modified Silica Gels

Silica gels of type 200-5-Si, (particle size: 5 μm, surface area S_{BET} = 200 m^2/g) were donated by the lab of Prof. K. Albert, University of Tübingen, Germany. The selectively deuterated C$_9$, C$_{18}$ and C$_{22}$-alkyl chains (CnSi-x, n = 9, 18 and 22 and x = 4, 6 and 12, see Figure 2.4) and non-deuterated n-alkyl modified silica gels were prepared by Singh et al.[21,23] using solution polymerization technique. The sample data are summarized in Table 4.2.

Table 4.2: Properties of various n-alkyl modified silica gels for studying the effects of pressure.

Sample	% C	Surface coverage (μmol/m^2)
C$_9$Si-4	9.2	5.2
C$_9$Si-6	6.6	3.5
C$_{18}$	14.3	4.2
C$_{18}$Si-4	13.8	4.0
C$_{18}$Si-6	13.9	4.1
C$_{18}$Si-12	14.1	4.2
C$_{22}$Si-4	14.8	3.6
C$_{22}$Si-6	18.4	4.8
C$_{22}$Si-12	22.0	6.2

4.1.4 C_{18}-Alkyl Modified Silica Gels with different Surface Coverages

C_{18}-alkyl modified silica gels with different surface coverages were prepared by Dr. Lane Sander, NIST, USA using 3 μm YMC silica with a surface area of 187 m^2/g. All phases were prepared with various amounts of octadecyltrichlorosilane and water to achieve stationary phases with a range of surface coverages. The essential parameters and properties are summarized in Table 4.3.

Table 4.3: Sample properties for studying the effect of surface coverage on C_{18}-alkyl modified silica gels.

Quantity water (ml)	Quantity silane (ml)	Quantity silica (g)	Reaction time (hr)	% C	Surface coverage (μmol/m²)	Selectivity Factor ($\alpha_{TBN/BaP}$)
0.5	1	4.1	4.5	7.3	2.0	1.22
anhydrous	10	4.0	18	9.4	2.7	1.59
0	10	3.9	5	14.4	4.5	0.94
0.5	10	4.0	5.5	15.7	5.0	0.71
humidified	10	4.0	25	22.3	8.2	0.25

4.1.5 C_{18} and C_{30}-Alkyl Modified Silica Gels

The procedures used for the polymeric phase synthesis have been described as "solution polymerized" and "surface polymerized" to distinguish the introduction of water for initiatiating the polymerization (Figure 2.2), and the description of these preparation techniques is described in Chapter 2.[16]

Three grams of silica gel were dried at 150 °C for 4 hours. The resulting dry powders were equilibrated with humid air for 60 min using a simplified apparatus. The humidified silica was then dispersed in 50 ml of xylene. 4.8 mmol of n-triacontyl trichlorosilane for C_{30} phases and 3.2 mmol of octadecyl trichlorosilane for C_{18} phases were dissolved in xylene and refluxed. The hot solution was filtered to remove solid impurities. The clear filtrate was then added to the slurry of silica. The temperature was kept at 110 to 120 °C and the mixture was stirred for 24 hours using a KPG stirrer to avoid breaking of the silica particles. The modified silica was filtered (G4 filter) and washed with acetone, ethanol, ethanol/water (1:1), water, ethanol, acetone and heptane sequentially. The filtered alkyl modified silica gel was dried under vacuum at 60 °C for 4 hours to remove all solvents used during washing. For solution polymerization, 1 ml water was added to the slurry of silica containing the triacontyl or octadecyl trichlorosilane and the aforementioned procedure was followed for the further steps. In the present work, the following samples are available, (i) C_{18} solution polymerized (hereafter denoted as C_{18} Solution), (ii) C_{18} surface polymerized (hereafter denoted as C_{18} Surface), (iii) C_{30} solution polymerized (hereafter denoted as C_{30} Solution) and (iv) C_{30} surface polymerized (hereafter denoted as C_{30} Surface). The parameters of the synthesized phases are given in Table 4.4.

Table 4.4: Sample properties for studying the effect of synthetic routes on C_{18} and C_{30}-alkyl modified silica gels.

Sample	% C	Surface coverage ($\mu mol/m^2$)
C_{18} Surface	16.78	5.21
C_{18} Solution	14.29	4.22
C_{30} Surface	20.86	4.07
C_{30} Solution	18.92	3.56

4.2 Elemental Analysis

Carbon and hydrogen analyses were performed on a Carlo Erba Strumentazione elemental Analyser 1106 (Italy). The percentage of carbon was utilized for calculating the surface coverage (α_{RP}) on the basis of the following equation,

$$\alpha_{RP} = \frac{10^6 P_C}{1200 n_C - P_C (M - n_x)} x \frac{1}{S_{BET}} [\mu mol/m^2] \qquad (4.1)$$

Here, P_C is the percentage of carbon determined via elemental analysis, n_C is the number of carbon atoms per silane moiety, M is the molar mass of the silane, n_x is the number of reactive groups in the silane ($n_x = 3$ for n-alkyltrichlorosilane), and S_{BET} is the specific surface area of the unmodified support. The silane loading on the silica surface was determined using the assumption as described in literature.[53]

4.3   Methods

4.3.1 FT IR Measurements

4.3.1.1   Sample Preparation and Variable Temperature Measurements

The samples were measured by means of the KBr pellet technique. The pellets of the alkyl modified systems and KBr (1/10 to 1/15 w/w) of 1 mm thickness were prepared under vacuum using a hydraulic press with a pressure of about 10 kbar. The pellets of the respective samples were placed in a brass cell equipped with an external thermocouple in close vicinity to the sample. The same thermocouple was also used for monitoring the actual sample temperature. The brass cell was thermostated in a variable temperature transmission cell (L. O. T. - Oriel GmbH, Langenberg, Germany) equipped with KBr windows. The temperature was regulated with an automatic temperature control unit with an accuracy of ±0.5 °C.

For pressure studies, the samples from deuterated and non-deuterated n-alkyl modified silica gels were prepared by two different methods. In method I, the powdered sample was placed between two KBr windows using a 25μm zinc spacer. In method II, the conventional KBr pellet technique method was used.

4.3.1.2   IR Measurements

IR spectra were recorded on a Nicolet Nexus 470 FT IR spectrometer with nitrogen purged optical bench (Nicolet, Madison, Wisconsin, USA) equipped with a DTGS detector. Typically, 256 interferograms covering a spectral range from 4000 to 400 cm^{-1} at a resolution of 2 cm^{-1} were collected within a temperature range from 193 to 353 K. The recorded interferograms were apodized with a triangular function and Fourier transformed with two levels of zero filling. Correction for background absorption was done by recording the background spectrum of the empty cell (measured with twice the number of interferograms as that used for the sample). The background spectrum was automatically subtracted from the spectra of respective samples. Data from three independent samples were acquired at all temperatures for all the samples studied and the FT IR spectra were measured twice for each sample.

4.3.1.3 IR Data Analysis

The processing and analysis of the spectra for CH_2 and CD_2 stretching band analysis was done with the OMNIC E.S.P.5.1 software (Nicolet, Madison, Wisconsin, USA). The frequencies of the CH_2 and CD_2 stretching vibrations were performed from the interpolated zero crossing in the first derivative spectra. Processing and analysis of the spectra in the CH_2 wagging regions were performed using the GRAMS 32 software (Galactic, Salem/New Hampshire, USA).

A quadratic baseline correction was applied in the spectral region from 1330 to 1400 cm^{-1}. The experimental spectra were fitted using four vibration bands. Their initial positions were 1378 cm^{-1} (symmetric methyl deformation mode), 1368 cm^{-1} (kink (gtg') and gtg sequences), 1354 cm^{-1} (double-gauche sequences) and 1341 cm^{-1} (end-gauche sequences). During the curve fit analysis, the band intensities and widths were varied independently. The integrated intensities of the CH_2 wagging bands were further normalized with respect to the methyl deformation band. The amount of specific gauche sequences was calculated according to the procedure given in literature, which is based on reference measurements on alkanes and a theoretical approach using the rotational isomeric state (RIS) model (see Spectroscopic Techniques).[82,83] Calculations reflect the fact that the surface linked alkyl chains contain only one methyl group per chain in contrast to liquid n-alkanes (which contain two methyl groups per chain). The total number of gauche conformers per chain was obtained by taking into account that two gauche bonds are necessary to form one kink or double-gauche sequence and one gauche bond for the formation of an end-gauche sequence. The estimated coefficient of variation for the various gauche conformers is 10% to 15%. In the case of C_{30} SAMs, a signal at 1384 cm^{-1} was observed for all samples. However, this transition has not been attributed to alkyl wagging or vibrational modes and is not considered in this analysis.

For the processing of CD_2 rocking bands, a quadratic baseline correction was applied for all samples. The CD_2 rocking band region comprises two bands, positioned at 651 cm^{-1} (gtgt, gtgt and ttgt) and 622 cm^{-1} (tt). The percentage of gauche bonds at a specific CD_2 segment can be calculated from the integrated intensities of these two bands by employing equation 3.7.[48]

4.3.2 NMR Measurements

Solid-state NMR experiments were carried out on a Bruker MSL 300 spectrometer operating at a static magnetic field of 7.05 T using a 4 mm magic angle spinning (MAS) probe. ^{13}C and

^{29}Si NMR experiments were done at 75.47 and 59.6 MHz, respectively. ^{13}C and ^{29}Si NMR spectra were recorded under MAS conditions (sample rotation frequency: 5 kHz) with cross-polarization (CP) excitation using a $\pi/2$ pulse width of 4.0 μs. A spin lock field of 62.5 kHz and contact time of 3 ms were employed during the CP experiments with recycle delays of 6 s. Typical numbers of scans were 2048. The ^{29}Si and ^{13}C chemical shifts were determined relative to external standards Q_8M_8 (trimethylsilylester of octameric silicate) and adamantane, respectively.

4.3.3 Chromatographic Measurements

Chromatographic measurements were carried out by Dr. Lane C. Sander, NIST for C_{18}-alkyl modified silica gels with surface coverages of 2.0 to 8.2 μmol/m^2 (see Chapter 4.1.4). Chromatographic columns were prepared by slurring packing the sorbents into 4.6 mm x 150 mm stainless steel tubes. Shape selectivity was assessed by use of Standard Reference Material (SRM) 869a Column Selectivity Test Mixture for Liquid Chromatography (NIST, Gaithersburg, MD).[88] The test was carried out with a mobile phase composition of 85:15 (volume fraction) acetonitrile:water, at a flow rate of 1.5 ml/min at ambient temperature (≈ 22 °C).

Chapter 5

Results and Discussion

In the present work, the influence of different solid supports, pressure, surface coverage, synthetic routes and solvents on the conformational order of the alkyl chains are examined with the help of FT IR and solid-state NMR spectroscopic techniques. The following sections present the experimental results and their comparison with published data on related systems, such as other n-alkyl modified silica gels, pure hydrocarbons, and biological membranes.

5.1   Effect of Solid Supports on C_{30} SAMs

The present work addresses the variable temperature FT IR studies that have been carried out to assess the conformational properties of C_{30} self-assembled monolayers attached to zirconia, titania and two different silica gels.[89] The C_{30} SAMs used in the present study were prepared by reacting $C_{30}H_{61}SiCl_3$ with the humidified surfaces of zirconia, titania, and two different silica gels (ProntoSil and LiChrospher). The essential parameters and properties of the C_{30} self-assembled monolayers are summarized in Table 4.1 (Experimental section).

Figure 5.1 illustrates the FT IR spectrum for the C_{30} SAMs in the anti-symmetric stretching band region between 2940 and 2900 cm^{-1}. The CH_2 anti-symmetric stretching band maximum varies by 10 cm^{-1} between the C_{30} SAMs on titania and the C_{30} SAMs on silica, which reflects significant differences in conformational order for these materials. The absorption band maxima of all samples shift towards higher wavenumbers with increasing temperature. Band maxima positions and variation with temperature are different for the various substrates. The

CH$_2$ stretching bandwidth increases with increasing sample temperature for C$_{30}$ SAMs on ProntoSil as reported earlier for n-alkanes and biomembranes (Figure 5.2).[81]

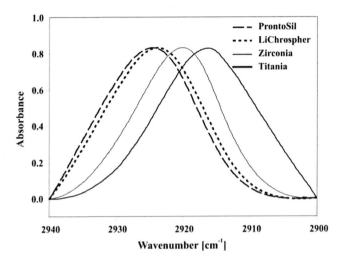

Figure 5.1: Experimental CH$_2$ stretching bands for C$_{30}$ SAMs on four different substrates at 353 K.

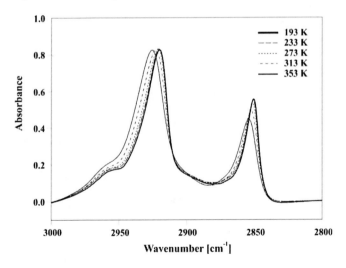

Figure 5.2: Temperature dependence of the CH$_2$ symmetric and anti-symmetric stretching modes of C$_{30}$ SAMs on silica (ProntoSil).

Absorption frequencies of the CH$_2$ symmetric and anti-symmetric stretching band regions are summarized in Figures 5.3 and 5.4. Inspection of these figures reveals that the frequencies of the symmetric and anti-symmetric stretching bands of the C$_{30}$ SAMs on titania are lower than for the other C$_{30}$ SAMs. Furthermore, the CH$_2$ symmetric and anti-symmetric stretching band positions for the C$_{30}$ SAMs on the two silica substrates are quite similar. Based on the assessment of the CH$_2$ stretching band positions, the conformational order of the C$_{30}$ SAMs can be ranked as titania > zirconia > silica, which holds over the whole temperature range.

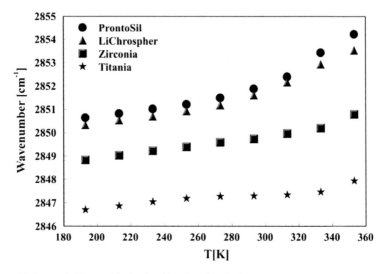

Figure 5.3: Symmetric CH$_2$ stretching band positions in various C$_{30}$ SAMs systems.

The conformation-dependent wagging modes of interest appear near 1368 cm^{-1}, 1353 cm^{-1} and 1341 cm^{-1} which arise from kink/gtg, double-gauche and end-gauche sequences, respectively, as discussed in Chapter 3. The most intense band in the spectral range from 1300 to 1400 cm^{-1} is due to the methyl group umbrella deformation mode at 1378 cm^{-1}, which is insensitive to the conformational order and is used as an internal reference (see Spectroscopic techniques). Figure 5.5 portrays typical FT IR spectra (only CH$_2$ wagging band region) of C$_{30}$ SAMs on ProntoSil at different temperatures from which the temperature dependence of the wagging band intensities is clearly evident.

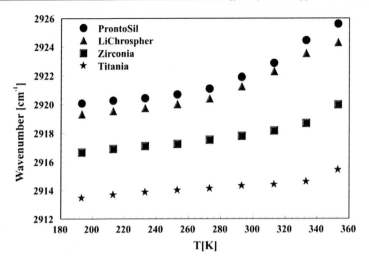

Figure 5.4: Anti-symmetric CH₂ stretching band positions in various C₃₀ SAMs systems.

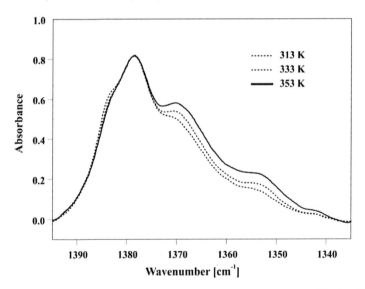

Figure 5.5: Experimental FT IR spectra (CH₂ wagging band region) for C₃₀ SAMs on ProntoSil at three different temperatures.

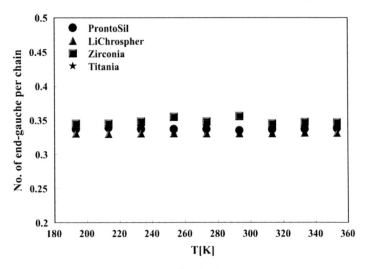

Figure 5.6: Number of end-gauche conformers per chain in the C$_{30}$ SAMs.

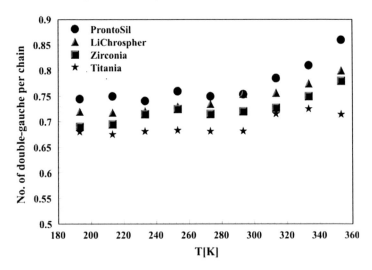

Figure 5.7: Number of double-gauche conformers per chain in the C$_{30}$ SAMs.

The results obtained from the curve fitting analysis of the CH$_2$ wagging bands are shown in Figures 5.6 to 5.8. The number of end-gauche conformers per chain remains almost unaffected by the sample temperature for all C$_{30}$ SAMs examined. In contrast, the number of kink/gtg and double-gauche conformers per chain is strongly temperature dependent for all C$_{30}$ SAMs. The most pronounced changes with temperature are observed for the kink/gtg conformers, particularly above room temperature. For the C_{30} SAMs on ProntoSil, the number of kink/gtg conformers per chain is between 0.68 and 0.87 at temperatures between 193 and 353 K, whereas for C_{30} SAMs on titania, the values lie between 0.53 and 0.67 over the same temperature interval. For LiChrospher, the number of kink/gtg conformers per chain varies between 0.65 and 0.85. In general, at low temperatures the fraction of kink/gtg conformers per chain is lower than the corresponding values for double-gauche conformers.

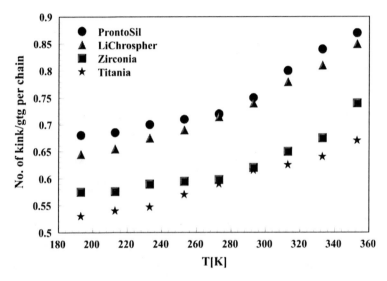

Figure 5.8: Number of kink/gtg conformers per chain in the C_{30} SAMs.

The total number of gauche conformers per chain for the four different C_{30} SAMs is depicted in Figure 5.9. As expected, the total number of gauche conformers per chain increases with temperature for all C_{30} SAMs. The higher alkyl chain disorder is observed for C_{30} SAMs on ProntoSil, with values between 3.1 and 3.7 gauche conformers per chain. The fewest defects per chain are found for C_{30} SAMs on titania, with values between 2.8 and 3.1. Inspection of Figure 5.9 further reveals that the effect of temperature is less pronounced for C_{30} SAMs on titania than on silica (ProntoSil and LiChrospher), which is similar to the trends observed for the CH_2 stretching frequencies.

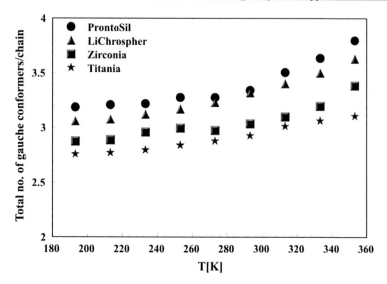

Figure 5.9: Variation of total number of gauche conformers per chain as a function of temperature in C_{30} SAMs.

The influences of both the sample type (i.e., substrate composition and/or C_{30} surface coverage) and the sample temperature are clearly evident from the present CH_2 stretching and wagging band data. In the case of the CH_2 stretching bands, a pronounced shift of the absorption maxima toward higher wavenumbers is visible upon increasing sample temperature. Although the CH_2 stretching data can only be discussed on a qualitative basis, the frequency shifts in CH_2 stretching bands and the changes in line broadening that are observed with increasing temperature are consistent with decreases in the conformational order. This stretching band data can be compared directly with the data from a recent FT IR study on various n-alkyl modified silica gels by Singh et al.[23] In that study – which included investigations on deuterated and non-deuterated alkyl chains – it was shown that the conformational disorder in the n-alkyl moieties is very sensitive to the temperature, proximity of particular CH_2 segments to the silica surface and the length of the n-alkyl chains. For instance, it was demonstrated that the conformational disorder in shorter n-alkyl chains (C_8 to C_{12}) is much greater than in their longer counterparts (C_{22} to C_{30}). A further comparison of the present CH_2 stretching band data with those from the C_{30}-alkyl modified silica gels[23] prepared via solution polymerization[17] reveals almost identical data for all silica supports, despite the differences in surface coverage and synthetic routes. It should be noted that the resemblance of these data for the various silica gels also includes the pronounced temperature dependence at elevated temperatures.

Another comparison of the present CH_2 stretching data can be made with the published data of hexadecanethiolate monolayers on a gold surface, in which the authors concluded that the number of gauche defects in the hexadecanethiolate chains is relatively low.[90] In a related study, Porter et al.[91] utilized reflection absorption infrared spectroscopy (RAIR) to study a wide range of alkanethiolate monolayers (chain lengths: C_2 to C_{23}) on a 2D gold surface. They compared CH_2 symmetric and anti-symmetric stretching for the alkanethiolate monolayers with those from crystalline hexadecanethiol (2918 cm^{-1}, 2851 cm^{-1}) and liquid heptanethiol (2924 cm^{-1}, 2855 cm^{-1}). The authors concluded that monolayers with chains longer than hexanethiol are highly ordered, whereas the shorter chain lengths resemble the liquid state with a greater fraction of gauche defects. Our present CH_2 stretching data indicates an even higher conformational alkyl chain order for the C_{30} SAMs compared with these former studies on alkanethiolate monolayers. These differences can probably be attributed to the longer chain lengths of the SAMs systems used in the present work.

Fadeev et al.[92] studied a series of $C_{18}H_{37}SiH_3$ SAMs prepared on various transition metal oxides by FT IR spectroscopy. They observed anti-symmetric stretching bands between 2916 and 2919 cm^{-1} for monolayers on TiO_2, ZrO_2 and HfO_2, which is consistent with ordering of the organosilicon hydride chains. By comparison, CH_2 stretching band values for C_{30} SAMs in the current work indicate an even higher degree of conformational order.

A distinct increase of the IR bandwidth with sample temperature is observed for all C_{30} SAMs, irrespective of the actual solid support. Such behavior is in qualitative agreement with the data reported by Griffiths et al.[93] for alkali salts of ascorbyl palmitate. Below 313 K, the CH_2 stretching bands are narrower indicating low acyl chain mobility. In addition, the absolute peak wavenumbers are characteristic for almost fully extended all-trans chains. The increase of the CH_2 stretching bandwidth with temperature is attributed to an increase in alkyl chain mobility along with a decrease in conformational order. Quite similar bandwidth changes were also reported from a FT IR study on phospholipid membranes.[81]

Quantitative information about alkyl ordering can be derived from analysis of the CH_2 wagging bands within the interval 1330 to 1400 cm^{-1}. Bands attributed to specific gauche defect sequences were obtained through deconvolution of the spectra over this interval using a four band model. Details of the approach have been described in the Chapter 3 (see

Spectroscopic techniques). In general, trends observed from the (quantitative) wagging band analysis are in agreement with the (qualitative) data from the CH_2 stretching bands.

The number of end-gauche conformers per chain was found to be nearly independent of temperature (and solid support), while a distinct temperature dependence is apparent for double-gauche and kink/gtg conformers. These observations for the C_{30} SAMs are consistent with the RIS model[82] developed for liquid alkanes, which predicts that the number of end-gauche conformers per chain is independent of the alkyl chain length, almost independent of temperature, and much lower than for the double-gauche and kink/gtg conformers. These predictions are consistent with a previous IR investigation on (long-chain) n-alkanes,[94] for which some evidence exists of nonlinear spectral changes in the vicinity of the melting point of the corresponding bulk alkane. This change reflects an increase in the number of non-planar (gauche) conformers as the melting point is approached.

The wagging mode band at 1368 cm^{-1} deserves additional discussion. Senak et al.[83] and Snyder[37] noted that this wagging mode arises from gtg' and gtg sequences within alkyl chains. The former sequence (gtg') is often referred to as a "kink" conformation whereas the gtg sequence represents a more bent conformation (see Figure 3.4). Because the relative contributions of the two conformers cannot be evaluated from the 1368 cm^{-1} wagging mode, this band cannot be used rigorously to quantify kink conformers alone. Senak et al.[83] reasoned that kink conformers probably dominate over gtg conformers due to chain packing considerations in alkyl liquids. For tethered alkyl chains (e.g., SAMs), spatial requirements would further restrict formation of gtg conformers. Kink/gtg conformers in the present C_{30} SAMs exhibit the strongest temperature dependence of the gauche defects studied. As a result, the temperature dependence of the combined gauche conformers per chain is dominated by the formation of the kink/gtg conformers.

It is interesting to examine differences in the influence of temperature on conformational defects for the four samples (Figures 5.6 to 5.8). Because the SAMs were prepared in the same manner for the four substrates, it might be expected that the materials would exhibit similar alkyl organization at the surfaces. Clearly, however, differences exist in the numbers of gauche defects per chain and temperature dependence of the alkyl conformations. The two silica based SAMs exhibit similar properties in terms of kink/gtg defects, even though the surface coverages differ significantly. The zirconia and titania SAMs represent more ordered systems. The temperature dependence of the double-gauche and kink/gtg conformers is somewhat different for the titania SAMs than for the other three materials, which suggests that a different organization of the alkyl chains exists on the substrate surface. In this connection, the question might arise whether the observed differences in chain conformational order are a consequence of the surface coverage. On the basis of the corresponding data in Table 4.1 such a conclusion cannot be drawn for the present systems. Another important contribution to the packing density of the alkyl chains and thus the alkyl conformational order is the uniformity of the surface coverage, i.e., whether the attached alkyl chains are clustered or whether they are distributed uniformly across the solid surface. The same holds for the constitution of the solid surface (roughness, specific surface area, porosity, etc.), which may also affect the alkyl chain properties. Unfortunately, there are not enough experimental data available in order to discuss the influence of the latter two contributions in an appropriate way.

The end-gauche conformational data in Figure 5.6 are consistent with the results reported by Singh et al.[23] for n-alkyl modified silica gels. They observed that the number of end-gauche conformers per chain was 0.35, irrespective of chain length and temperature. In addition, the influence of temperature on the number of double-gauche conformers per chain was small for the C_{30}-alkyl modified silica gels, which was attributed to the higher chain packing and lower conformational disorder in these long alkyl chains. For shorter alkyl phases, the number of double-gauche conformers and their temperature dependence was found to be much more pronounced. In the current work, elevated temperature was found to influence gauche conformational defects to a greater extent than subambient temperature (see Figures 5.7 and 5.8).

The data for the various amounts of gauche conformers – as evaluated from the CH_2 wagging band data – can be further compared with previous investigations on phospholipids, where such FT IR techniques have been employed extensively. Thus, the liquid crystalline phases of phospholipid bilayers from 1,2- dipalmitoyl-phosphatidylcholine (DPPC), 1,2-dimyristoyl-phosphatidylcholine (DMPC), 1,2- dipalmitoyl-phosphatidylethanolamine (DPPE)[83,95] are characterized by a lower fraction of double-gauche conformers, while end-gauche and kink/gtg conformers are very similar to the values determined for the present C_{30} SAMs. The total number of gauche conformers – and thus the conformational disorder – is generally lower in DPPC, DMPC and DPPE bilayers than in C_{30} SAMs, although the alkyl chain lengths in these systems are quite different (i.e., C_{14} or C_{16} in the phospholipids vs. C_{30} in the SAMs).

Titania and silica based C_{30} SAMs have been examined by Pursch et al.[20] using LC, atomic force microscopy, ellipsometry, and solid-state ^{13}C NMR spectroscopy. Solid-state ^{13}C NMR spectroscopy was used to derive information about the conformational order of the attached alkyl chains by measuring the relative intensities of carbon signals attributed to trans and gauche conformations. Two distinct signals for the inner methylene groups of the alkyl chain were observed for C_{30} SAMs on silica (LiChrospher). The chemical shift for the trans peak of the C_{30} SAMs (33.4 ppm) is nearly identical to that of the interior methylene carbons of long-chain alkanes in the crystalline state, while the gauche peak at 30.6 ppm agrees with the chemical shift value in the isotropic alkane melt. A similar signal splitting of these methylene resonances were reported for C_{30}-alkyl modified silica gels in which the interconversion of the two conformational states was monitored. At elevated temperatures a larger fraction of gauche conformers was detected.

It should be emphasized that the results from the present FT IR study are consistent with the ^{13}C NMR data by Pursch et al.[20] on the same systems, where also higher conformational order for the titania and zirconia SAMs was reported as compared to the silica SAMs. In addition, the ^{13}C NMR line width of the trans peak for the titania SAMs was reported to be approximately three times larger than for the corresponding interior methylene resonance in crystalline $C_{19}H_{40}$ and $C_{32}H_{66}$, which was attributed to a general higher disorder (i.e., greater chemical shift dispersion) as well as a higher mobility – causing line broadening – in the SAMs as compared with crystalline alkane phases.

In principle, the existence of all-trans chains should be visible in FT IR spectroscopy via the formation of wagging band progressions[48] in the region between 1350 and 1100 cm^{-1}. This phenomenon has been reported for pure hydrocarbons[94] in the crystalline state and phospholipid membranes[96,97] in the gel state. In the present study on the C_{30} SAMs such wagging band progressions were not detected. It is likely that sufficient numbers of gauche conformers exist in these materials to prevent the formation of progression bands.

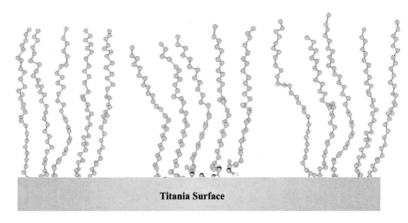

Figure 5.10: Schematic picture of C_{30} SAMs on titania.

In summary, from the analysis of both the CH_2 stretching and CH_2 wagging bands it was found that alkyl chain flexibility and conformational disorder increases with increasing sample temperature. The highest degree of conformational order exists for SAMs prepared on titania, followed by zirconia and by the silica supports. From an analysis of the wagging bands, the number of kink, double-gauche, end-gauche, and total number of gauche conformers per chain were calculated. It is shown that the observed temperature dependence of the conformational order is dominated by the changes in the number of kink/gtg conformers. The molecular origin of the somewhat higher conformational order for the titania support is so far unknown. It certainly cannot be related to the surface coverage, as both the zirconia and the titania supports exhibit a very similar surface coverage. It might be speculated whether the titania sample exhibits a higher surface inhomogeneity, expressed by areas of local high bonding density referred as "islands" (see Figure 5.10). This would then give rise to a higher chain packing density in these areas along with the observed higher conformational order. However, so far there is no direct proof for this assumption. Furthermore, the question might then arise, why a similar surface inhomogeneity does not exist for the other solid supports as well. It is therefore quite obvious that much more work is necessary in order to provide a satisfactory explanation of the derived alkyl chain order in the systems examined here.

5.2   Effect of Solid Supports on C_{18} Systems

In the following, FT IR and solid-state ^{13}C NMR investigations probing the effect of temperature, surface coverage and solid supports on the conformational order and mobility of the alkyl chains of commercially available DiamondBond®-C_{18} systems in the dry state are presented. In the present work two different DiamondBond®-C_{18} column materials are studied, i.e., (i) low density DiamondBond®-C_{18} (hereafter denoted as LDZr-C_{18}) with a surface coverage of 2.5 μmol/m², and (ii) high density DiamondBond®-C_{18} (hereafter denoted as HDZr-C_{18}) with a surface coverage of 4.0 μmol/m². The results are directly compared with the data from a parallel study on octadecyl-modified silica gels with a surface coverage of 4.2 μmol/m² (denoted as Si-C_{18}) and with data on related systems.[98]

To understand the local molecular environments and structural arrangement of the carbon-clad zirconia based DiamondBond®-C_{18} systems and C_{18}-alkyl modified silica gels (see Figure 2.4), variable temperature FT IR as well as NMR spectroscopy are utilized, since both can be used to get complementary information of the tethered alkyl chains attached to the different metal oxides. Conformational order in the DiamondBond®-C_{18} column material and C_{18}-alkyl modified silica gel are assessed using FT IR spectroscopy from the position of the CH_2 symmetric and anti-symmetric stretching bands. In principle, quantitative information about the presence and amounts of various gauche conformers (kink/gauche-trans-gauche (gtg), double-gauche, and end-gauche conformers) is feasible by the analysis of the wagging band intensities between 1330 and 1400 cm^{-1}. The analysis was not possible for the present DiamondBond®-C_{18} systems due to overlapping of other peaks, which are likely due to the phenyl-graphite layer modified substrate. Therefore, only qualitative studies are performed for the present column materials. Several attempts are made in order to study the conformational order of DiamondBond®-C_{18} systems in the presence of various mobile phases, which would be of more interest for the comparison under chromatographic conditions. In fact, it is known that the presence of a mobile phase has an impact on the conformational properties of the attached alkyl chains as compared to the dry state.[25] In order to do so, the DiamondBond®-C_{18} powdered samples were placed between two KBr windows. Then solvent was introduced into the KBr windows to study the influence of solvents on the conformational order of alkyl chains in the DiamondBond®-C_{18} systems. In this case, however, the spectral quality was so poor, most probably due to the higher concentration of DiamondBond®-C_{18} materials, which made the analysis of the IR spectra impossible. Several attempts to overcome this problem were not successful. Thus, the present FT IR study only comprises experiments of dry materials.

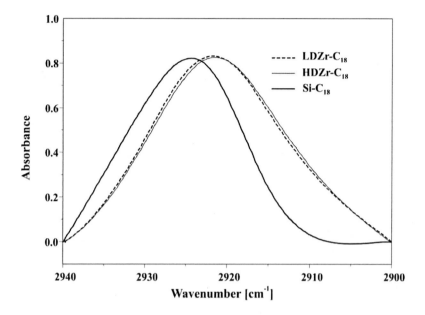

Figure 5.11: Experimental anti-symmetric CH_2 stretching bands for LDZr-C_{18}, HDZr-C_{18} and Si-C_{18} phases at 333 K.

Figure 5.11 portrays the FT IR spectrum for the HDZr-C_{18}, LDZr-C_{18} and Si-C_{18} in the anti-symmetric region between 2940 and 2900 cm^{-1}. The CH_2 anti-symmetric band maximum for both DiamondBond$^{®}$-C_{18} phases is almost the same, whereas the band maximum varies by 4 cm^{-1} between silica based and zirconia based supports, i.e., DiamondBond$^{®}$-C_{18} phases. Upon increasing the sample temperature the corresponding absorption band maxima of the samples shift towards higher wave numbers independent of the surface coverage. The position of the band maxima as well as the actual variation with temperature, however, depends on the respective samples. Figure 5.12 depicts FT IR spectra for the HDZr-C_{18} phases in the symmetric and anti-symmetric stretching band region between 3000 and 2800 cm^{-1}. The impact of temperature on the CH_2 symmetric and anti-symmetric stretching band maxima of the DiamondBond$^{®}$-C_{18} systems is found to be lower as compared to conventional C_{18}-alkyl modified silica gels column materials.

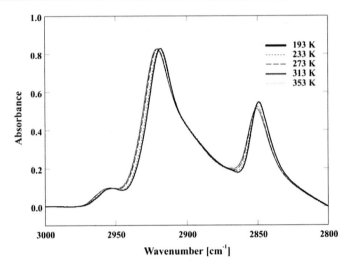

Figure 5.12: Temperature dependence of the symmetric and anti-symmetric CH_2 stretching bands of HDZr-C_{18} phases.

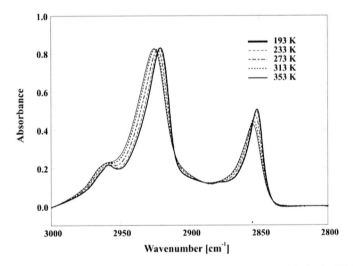

Figure 5.13: Temperature dependence of the symmetric and anti-symmetric CH_2 stretching bands of Si-C_{18} phases.

It is important to note that the increase of FT IR bandwidth with increasing temperature -a phenomenon described in the earlier studies[81,89,93] and found for the present Si-C_{18} sample - is absent in the DiamondBond®-C_{18} systems. In the case of Si-C_{18} (Figure 5.13), the CH_2 bands are narrow at lower temperatures, reflecting a low alkyl chain flexibility and high conformational order. At higher temperatures, an increase of the CH_2 stretching bandwidth is observed which can be attributed to an enhanced alkyl chain flexibility owing to a decrease of conformational order.

The derived results for the absorption frequencies of the CH_2 anti-symmetric and symmetric stretching band regions of the samples are summarized in Figures 5.14 and 5.15. Inspections of these figures reveal that the wavenumbers of the symmetric and anti-symmetric stretching bands of the LDZr-C₁₈, HDZr-C₁₈ phases are lower as compared to the Si-C₁₈ phases for the entire temperature range. These stretching band data imply that DiamondBond®-C₁₈ column materials exhibit a higher conformational order as compared to the C₁₈-alkyl modified silica gels systems. Moreover, the impact of temperature on the silica-based systems is more pronounced as compared to the zirconia-based DiamondBond®-C₁₈ systems, which maintain their high conformational order over a wider temperature range. The existence of higher conformational order, in particular for an extended temperature range, is considered to be the molecular basis of the enhanced performance of the DiamondBond®-C₁₈ column materials during chromatographic separations.[99,100]

In comparing the high and low surface coverage DiamondBond®-C₁₈, the HDZr-C₁₈ has a slightly higher conformational order as compared to the LDZr-C₁₈ systems. The surface coverage has a greater influence at lower temperatures than at higher temperatures. As expected, these results indicate that surface coverage has a direct influence on the conformational order of the alkyl chains. Even though the surface coverage is less in DiamondBond®-C₁₈ systems ($2.5 \mu mol/m^2$ for LDZr-C₁₈, $4 \mu mol/m^2$ for HDZr-C₁₈) as compared to Si-C₁₈ ($4.2 \mu mol/m^2$), the influence of the solid supports dominates over the influence of surface coverage, as can be deduced from the present results. It should be emphasized that the C₁₈-alkyl modified silica gels and DiamondBond®-C₁₈ systems are chemically distinct from each other. As shown in Figure 2.4, the zirconia particle surface is clad with a graphite-like layer linked with phenyl groups followed by tethered C₁₈ alkyl chains, whereas the alkyl chains are directly connected to the surface silanol groups of the silica surface in the case of C₁₈-alkyl modified silica gel.

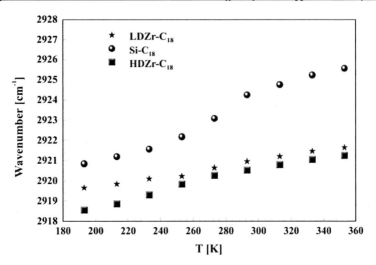

Figure 5.14: Anti-symmetric CH_2 stretching band positions of LDZr-C_{18}, HDZr-C_{18} and Si-C_{18} phases.

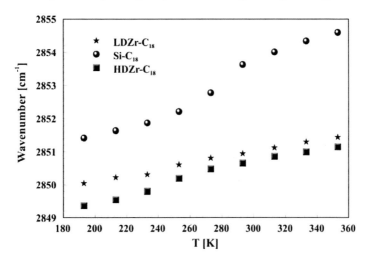

Figure 5.15: Symmetric CH_2 stretching band positions of LDZr-C_{18}, HDZr-C_{18} and Si-C_{18} phases.

Higher conformational order is observed for carbon clad zirconia-based systems when compared with silica based modified silica gels. One might be tempted to argue whether the ordering of alkyl chain originates from the zirconia supports or it stems from the carbon clad (graphite materials) that is coated on the surface of the zirconia supports. At present, a definitive answer on this issue cannot be given. The present findings, however, are consistent with previous data for C_{30} SAMs on different solid supports (see Chapter 5.1).[89] Moreover, like in the present DiamondBond®-C_{18} systems, a reduced temperature dependence of the alkyl chain conformations was observed for the C_{30} SAMs on zirconia and titania.

In the present systems an increase of conformational order is observed upon increase of surface coverage. This is consistent with the data from a comprehensive study of surface coverage on alkyl modified silica gels (see Chapter 5.4). Likewise, Fadeev et al.[92] studied a series of zirconia-based samples that reacted with octadecylsilane over different periods of time. A gradual shift towards lower wave numbers was observed indicating a transition from completely disordered to ordered structures as the surface coverage of the monolayers increases.

One might wonder whether the CH_2 stretching band differences between the zirconia and silica based materials are due to the effect the carbon layer might have on the spectra and not a result of increased conformational order. However, earlier investigations on SAMs (see Chapter 5.1), biomembranes, alkanes and polymers[37,48,92,89] proved that the CH_2 stretching band positions are irrespective of the corresponding materials, and always occur in the same spectral range, as discussed earlier. Independent results from NMR[4,19] and Raman spectroscopy[25-28] as well as studies of other conformational sensitive IR bands clearly proved that the variation of the stretching frequencies is only due to changes in the conformational order. Remarkable changes in the CH_2 stretching absorption would be only expected if the respective support and the tethered alkyl chains strongly interact, for which there is no evidence in the present systems.

Variable temperature ^{13}C CP/MAS NMR spectra of the LDZr-C_{18} material are depicted in Figure 5.16. The ^{13}C NMR chemical shifts for $(CH_2)_n$ units of DiamondBond®-C_{18} is observed at 30.2 ppm, which is attributed to gauche defects. The peak at 23.5 ppm characterizes the methylene unit attached to the phenyl group, and the resonance at 13 ppm shows the presence of terminal methyl group. The resonance at 129.5 ppm and the other resonances above 129.5 ppm occur due to the presence of the phenyl group and the graphite layers in the DiamondBond®-C_{18} material (Figure 2.4). The main signal for the $(CH_2)_n$ groups is almost unaffected by the actual sample temperature.

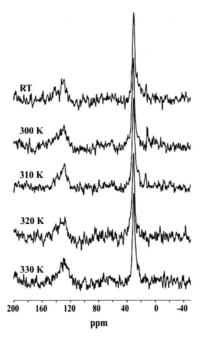

Figure 5.16: Variable temperature ^{13}C CP/MAS NMR spectra for the LDZr-C$_{18}$ phases.

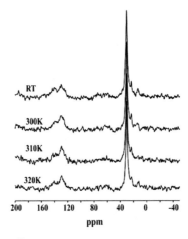

Figure 5.17: Variable temperature ^{13}C CP/MAS NMR spectra for the HDZr-C$_{18}$ phases.

The ^{13}C CP/MAS NMR spectra of the HDZr-C$_{18}$ material are shown in Figure 5.17, where the main resonance for the (CH$_2$)$_n$ is observed at 30.5 ppm, which is attributed to the disordered chain segments (liquid-like equilibrium of trans/gauche conformations). In comparison with the LDZr-C$_{18}$ material, it is seen that the resonance frequency is slightly lowfield shifted, which suggests a reduced amount of disordered chain segments. Upon increase of temperature, the main resonance for the (CH$_2$)$_n$ is again practically unaffected. That is, the ^{13}C NMR data and the aforementioned FT IR data for the DiamondBond$^{®}$-C$_{18}$ samples provide consistent results with respect to the influence of temperature on the alkyl chain conformational order.

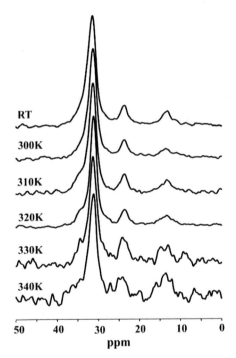

Figure 5.18: Variable temperature ^{13}C CP/MAS NMR spectra for the Si-C$_{18}$ phases.

The variable temperature ^{13}C CP/MAS NMR spectra of the Si-C$_{18}$ (solution polymerized) are shown in Figure 5.18. For the Si-C$_{18}$ phases at room temperature, the NMR resonances for the main (CH$_2$)$_n$ resonance appear at 31.4 ppm, and at 13.2 ppm for the terminal CH$_3$ group. The value of 31.4 ppm for the main methylene resonance can be attributed to trans-gauche chains, which is in between the literature values for all-trans chains (~ 33 ppm) in crystalline alkanes and conformationally disordered gauche chains (~30 ppm) in isotropic melts.[101] For the present C_{18} chains two distinct methylene group signals, referring to these two types of alkyl chains of different conformational state, could not be observed simultaneously. Typically, such separated peaks only show up for longer alkyl chain lengths, such as C_{22}, C_{30} or C_{34}.[20,22,56] Figure 5.18 further demonstrates that the amount of gauche conformations increases with increasing temperatures, since the main resonance shifts slightly upfield with a value of 31.0 ppm at 340 K. At higher temperatures, dual methyl resonance profiles are obtained at chemical shift values of 14.7 and 13 ppm.

On comparing the position of the main NMR resonances of the $(CH_2)_n$ groups in the DiamondBond®-C_{18} phases and C_{18}-alkyl modified silica gels, i.e., upfield shifts in the DiamondBond®-C_{18} phases, one might be tempted to assign a "less ordered" organization to the methylene carbons $(CH_2)_n$ of DiamondBond®-C_{18} phases and conclude that the DiamondBond®-C_{18} phases have higher amounts of gauche concentrations. However, such upfield shifts have been observed for ^1H NMR resonances of adsorbates (chemisorbed) on other carbonaceous materials such as water on cokes,[102] xenon on grafoil[103] and a variety of molecules adsorbed on the basal plane of graphite carbon black.[104] These findings were related to the existence of a highly anisotropic magnetic susceptibility of the ^{13}C CP/MAS NMR powder pattern in graphite and related materials. As a result, the alkyl chains attached to the graphitic surface via phenyl group experience a strong diamagnetic field component due to the anisotropic magnetic susceptibility of the graphite layers.[105] Thus, the present overall upfield shift of the ^{13}C resonances of the DiamondBond®-C_{18} phases does not reflect directly the conformational order. Rather it stems from the magnetic interaction with the graphite surface.

Therefore it is not possible to compare the ^{13}C CP/MAS NMR spectra of the DiamondBond®-C_{18} phases and the conventional C_{18}-alkyl modified silica gels directly. Obviously, for the present systems the IR spectroscopic data, i.e., CH_2 stretching band positions, are more reliable. As mentioned earlier, the temperature dependencies of the ^{13}C resonances and CH_2 stretching band position for a distinct sample, however, are consistent.

Emphasis is also given to the terminal methyl group to understand the behavior of alkyl chain structure especially on its surface. Dual methyl resonance peaks are observed only for silica based systems especially at higher temperatures i.e., above 330 K. Pursch et al. discussed the methyl group resonances in C_{30} SAMs in detail as an important characterization tool for these alkane-type SAMs.[20] Here, for C_{30} SAMs on ProntoSil (silica gel), dual methyl resonance profiles were observed at chemical shifts of 14.7 and ~ 13 ppm. A similar upfield shift has been noted for chain-end methyl groups in polyethylene when going from a crystalline to a noncrystalline environment.[106] The split up of the methyl resonance was attributed to the heterogeneity of the sample, i.e., methyl groups of different mobilities exists in these materials.

In summary, solid-state ^{13}C NMR and FT IR spectroscopy for two different C_{18} surface coverages DiamondBond®-C_{18} phases and conventional C_{18} phases are examined. The temperature dependence of the ^{13}C NMR and FT IR data was found to be consistent. However, contradictory results were obtained from the FT IR and ^{13}C NMR data with respect to the influence of the solid support, which can be attributed to the existence of a highly anisotropic magnetic susceptibility of carbon cladding which is vapour deposited on the zirconia substrate. We conclude that the HDZr-C_{18} has a greater conformational order as compared to its corresponding low density counterpart. The temperature dependence of the conformational order of the C_{18} chains on the carbon clad zirconia phase is much less pronounced than in conventional column material prepared from silane modified silica gels. It is concluded that carbon clad zirconia based substrates exhibit a higher conformational order of alkyl chains and high thermal stability, which should be the molecular origin of the enhanced performance of the DiamondBond®-C_{18} systems during chromatographic separations.

5.3   Effect of Pressure

In the following, a comprehensive variable temperature FT IR study on non-deuterated C_{18} chains as well as specifically deuterated C_9, C_{18} and C_{22} chains (deuterated positions: carbons C-4, C-6, C-12, samples CnSi-x, n = 9, 18 and 22 and x = 4, 6 and 12, see Figure 2.4) in the dry state are presented. Particular emphasis is given to the effect of the sample preparation method, i.e., sample pressure, on the conformational order of the tethered alkyl chains. Thus, two series of samples have been examined: (i) samples which were prepared at normal pressure (method I), and (ii) samples which experienced a pressure of about 10 kbar during preparation (KBr pellet technique, method II).[107]

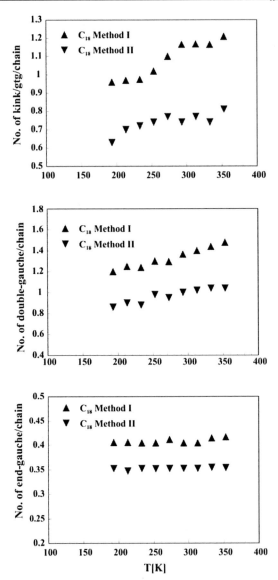

Figure 5.19: Variation of kink/gtg, double-gauche and end-gauche conformers as a function of sample preparation method and temperature.

The FT IR spectra for n-octadecyl (C_{18}) modified silica gels, with emphasis on the CH_2 wagging band region, were recorded throughout a temperature range from 193 and 353 K.

The results of the CH_2 wagging band analysis for both n-octadecyl modified silica gel samples, which differ by their sample preparation method, are summarized in Figure 5.19.

An inspection of these graphs reveals a substantial increase of the amount of kink/gtg, double-gauche conformers per chain with increasing temperature, which holds for both preparation methods. The temperature dependence of the kink/gtg conformers, especially above the room temperature, is somewhat more pronounced for the sample from preparation method I. Moreover, for the same sample the amount of kink/gtg conformers with values between 0.95 and 1.2 is lower than the corresponding values of 1.2 to 1.5 for the double-gauche conformers. The amount of end-gauche conformers per chain (about 0.4) remains almost unaffected by the actual sample temperature. It is found that the present results for the various gauche conformers in n-alkyl modified silica gels are consistent with the RIS model developed for liquid alkanes, which predicts that the amount of end-gauche conformers is independent of the alkyl chain length and temperature, and lower than the double-gauche and kink/gtg conformers. The total number of gauche conformers per chain in the n-octadecyl modified silica gel is depicted in Figure 5.20. The total number of gauche conformers for method I varies between 4.7 and 5.7 in the temperature range covered here.

The impact of the actual sample preparation method, i.e., application of external pressure, on the conformational properties of n-octadecyl modified silica gels is clearly evident from the graphs given in Figures 5.19 and 5.20. It can be seen that an external pressure (method II) in general increases the conformational order, as reflected by the lower amount of gauche conformers per chain. The relative changes are largest for the kink/gtg and double-gauche conformers. Moreover, it is found that the temperature dependence also is less pronounced for the sample, which experienced a high external pressure.

Similar tendencies are registered during the analysis of the CD_2 stretching and rocking bands from selectively deuterated n-alkyl modified silica gels, which will be presented next. The absorption frequency of the CD_2 symmetric and anti-symmetric vibration - at 2115 to 2070 cm^{-1}, and 2200 to 2165 cm^{-1}, respectively - provides qualitative information about the conformational order at the labelled methylene segment. A shift of the band maxima toward higher or lower wavenumbers thus indicates a decrease or increase of conformational order in the system of interest, respectively.

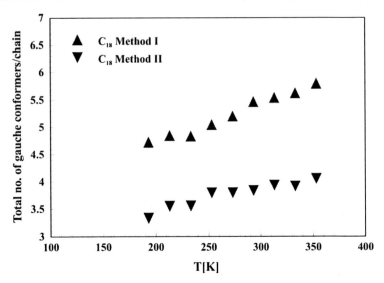

Figure 5.20: Variation of total number of gauche conformers as a function of sample preparation method and temperature.

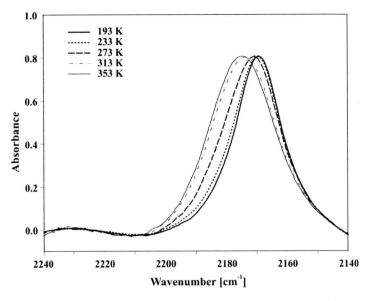

Figure 5.21: Temperature dependence of the anti-symmetric CD_2 stretching modes of C_{18} modified silica gels deuterated at the position 6 (C_{18}Si-6), prepared by method I.

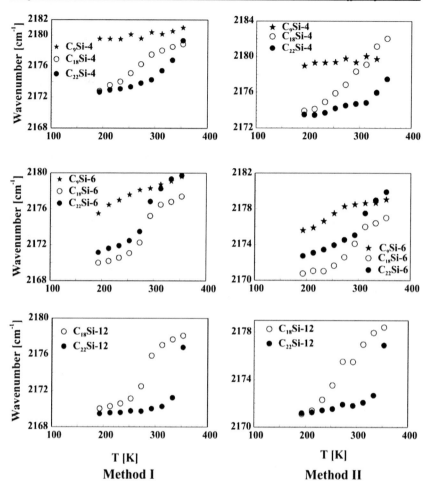

Figure 5.22: Anti-symmetric CD_2 stretching band positions as a function of temperature for the selectively deuterated n-nonyl, n-octadecyl and n-docosyl modified silica gels.

A representative series of anti-symmetric stretching bands from measurements on sample $C_{18}Si$-6 is depicted in Figure 5.21. The general observation is a shift of the corresponding absorption band maximum toward higher wavenumbers upon increase of the sample temperature, which points to an enhanced conformational disorder in the same direction, in close agreement with the earlier findings from the CH_2 wagging band analysis (see above). In addition, the shift of the absorption band maximum toward higher wavenumbers is accompanied by a increase of the IR bandwidth (see Figure 5.21), a phenomenon which also was discussed before during IR studies on n-alkanes, phospholipids (CH_2, CD_2 stretching bands)[81,93] and more recently in C_{30} SAMs (see Chapter 5.1).[89] At low temperature, the CD_2 stretching bands are narrow, reflecting relatively low acyl chain flexibility and high conformational order, whereas the increase of the CD_2 stretching bandwidth with temperature is attributed to an enhanced alkyl chain flexibility due to the decrease in conformational order.

The derived results from the analysis of the CD_2 anti-symmetric stretching bands are summarized in Figure 5.22. The various graphs clearly indicate that the conformational order critically depends on the actual chain length and chain position. For instance, the absorption maxima of the n-nonyl system occur at higher wavenumbers due to a higher conformational disorder as compared to the silica gel samples with longer alkyl chain lengths. In addition, it is registered that the impact of the pressure during sample preparation on the conformational order is less pronounced in the CD_2 stretching band data.

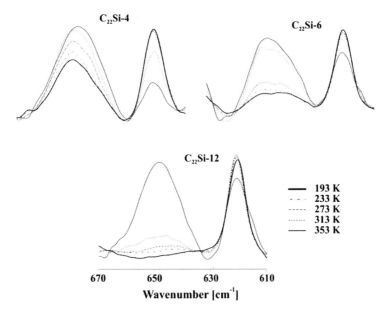

Figure 5.23: Variable temperature FT IR (CD_2 rocking band region) of n-docosyl alkyl modified silica gels, selectively deuterated at positions C-4, C-6 and C-12 prepared by method I.

Representative variable temperature FT IR spectra, covering the CD_2 rocking band region, for the C_{22}-alkyl modified silica gels labeled at the positions C-4, C-6 and C-12 of the alkyl chains are shown in Figure 5.23. These experimental spectra clearly demonstrate that the relative intensity of the vibration band at 651 cm^{-1}, characterizing the gauche conformers, is considerably higher for chain position C-4 than for positions C-6 and C-12. These experimental FT IR spectra again prove that the actual amount of gauche conformers depends on both the temperature and chain position. Moreover, the comparison with the spectra from the samples with octadecyl and nonyl chains (spectra not shown) reveals a distinct chain length dependence.

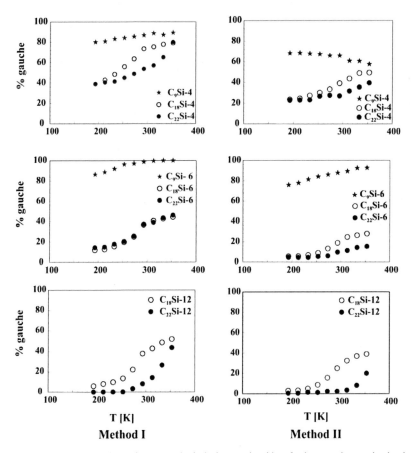

Figure 5.24: Amount of gauche conformers at selectively deuterated positions for the n-nonyl, n-octadecyl and n-docosyl modified silica gels as a function of temperature.

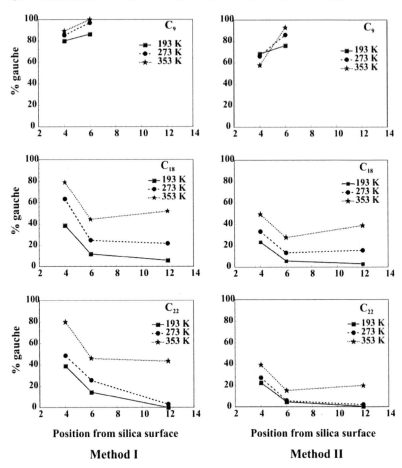

Method I   **Method II**

Figure 5.25: The position dependence of the gauche amounts in n-alkyl modified silica gels at 193 K, 273 K and 353 K.

The graphs in Figure 5.24 depict the amount of gauche conformers as a function of temperature, alkyl chain length and deuterated chain position for both sample preparation methods. In Figure 5.25 the percentage of gauche conformers is plotted against the chain position for all three n-alkyl chain lengths and for all deuterated chain positions at three selected temperatures (193, 273 and 353 K). For the octadecyl chains the amount of gauche conformers at position C-4 (sample $C_{18}Si$-4) varies from 38% to 80% for method I, while the corresponding sample from method II exhibits only 24% to 40% gauche conformers. In agreement with the former CH_2 wagging data, the effect of temperature on the conformational order of the tethered alkyl chains is more pronounced for preparation method I (i.e., without external pressure).

The conformational order of the system with the shortest alkyl chain length (C$_9$) is considerably lower than for its longer counterparts. Thus, for sample C$_9$Si-4 (method I) the amount of gauche conformers at position C-4 is already 80% even at low temperatures, while for the sample prepared at higher pressure (method II) 70% gauche conformers are found. The temperature dependence, however, is different. The usual increase upon sample heating is registered for method I (89% at 353 K), while for method II the amount of gauche conformers decreases to 57% at elevated temperatures.

Upon moving along the alkyl chains and away from the silica surface ($\rightarrow$ position C-6 $\rightarrow$ position C-12), the percentage of gauche conformers in the octadecyl chains gradually decreases at lower temperatures or goes through a minimum at position C-6 at higher sample temperatures (see Figure 5.24). The different amounts of gauche conformations for the chain positions C-6 and C-12 as a function of the external pressure is quite striking for both the octadecyl and docosyl modified silica gels (see Figures 5.24 and 5.25). That is, the conformational order considerably increases upon applying an external pressure during sample preparation. In general, the conformational order of the octadecyl and docosyl modified silica gels is found to be very similar, which holds for the chain position, sample temperature and pressure dependence, although a somewhat lower amount of gauche conformers is registered for the latter system. These findings are in qualitative agreement with the corresponding CD$_2$ stretching and CH$_2$ wagging band data (see above).

The highest amount of gauche conformers is found at position C-6 in the n-nonyl modified silica gels, where 86% gauche conformers exist at 193 K. This value increases even further upon raising the sample temperature. It should be emphasized once again that the derived amounts of gauche conformers from the CD$_2$ rocking band analysis are in qualitative agreement with the corresponding CD$_2$ stretching band data, which in particular holds for the position, chain length and temperature dependence, although the CD$_2$ stretching bands can only considered to be a qualitative measure of the conformational order.

On the basis of present variable temperature FT IR study we can thus conclude that the conformational order of the tethered alkyl chains in chemically modified silica gels

a) decreases upon raising the sample temperature,

b) increases by increasing the alkyl chain length,

c) increases gradually with increasing distance from the silica surface (or possesses a minimum at position C-6 at higher temperature), and

d) increases upon applying pressure during sample preparation.

The present results are in agreement with former spectroscopic studies on related systems.[21,23,89] Sander et al.[24] described the application of FT IR spectroscopy to the study of the conformational order of C_1, C_4, C_8, C_{12}, C_{18} and C_{22} alkyl modified silica gels by dealing with C-H stretching modes for methyl and methylene groups as well as C-H bending, scissoring and wagging modes. In these studies, a significant fraction of gauche conformers in the bonded n-alkyl chains was observed. Likewise, a Raman study on C_{18} modified silica gels from commercially available silica-based packing materials in the ν(C-C) and ν(C-H) spectral regions were performed.[108] Again, the experimental data suggested that the n-alkyl chains of these C_{18} modified stationary phases exist in a highly disordered state, which is similar to the case of (liquid) free n-alkanes. In our present work the C_{18} modified silica gel exhibits less conformationally disordered alkyl chains than in the corresponding free n-alkane, which most likely reflects the three-fold higher surface coverage than that in the work by Pemberton et al.[108] The surface coverage is certainly a crucial factor for the discussion of the molecular properties in such modified silica gels. A comprehensive study of the influence of surface coverage on the alkyl chain conformational order will be discussed in Chapter 5.4.

The conformational disorder in the high-temperature crystal phases of long n-alkanes ($C_{21}H_{44}$ and $C_{29}H_{60}$) were studied via CD_2 rocking vibrational modes.[109] It was found that the gauche conformers in these pure n-alkanes consisted entirely of kink conformations, distributed non-uniformly along the chain. The highest amount of gauche conformers was found at the chain ends and the concentration at interior sites decreased gradually on moving toward the center of the chain. In the present work, gauche conformers were absent for n-docosyl modified silica gels at position C-12 at temperatures below 250 K, which is thus comparable with the findings of crystalline $C_{21}H_{44}$ at this position. In addition, it is found that the amounts of gauche conformers at higher temperatures in the n-docosyl modified silica gels are very close to those reported for $C_{21}H_{44}$ in the liquid phase.[110] Silica gels are amorphous materials and a direct comparison with the crystalline n-alkanes should be done with caution; in particular, it should be mentioned that in the present silica gels a calorimetric phase transition is not observed. Nevertheless, the high conformational order at low temperatures in the inner part of the n-docosyl chains attached to the silica surface must reflect an extremely good chain packing of this system.

Chemically modified n-alkyl silica gels have conformational features similar to those of surface bound n-alkyl chains in SAMs. Transmission IR studies for n-alkanethiolate SAMs on gold clusters for various n-alkyl chain lengths (C_3 to C_{24}) have been performed by Murray et al.[111] The analysis in that work was based on the qualitative comparison of the intensity of a band, which depends on the n-alkyl chain length (e.g. CH_2 symmetric stretching, CH_2 wagging) with that of a band, which is invariant of the chain length (e.g. symmetric CH_3 stretching mode). Again, it was shown that for such systems shorter n-alkyl chains (C_3 to C_5) are highly disordered and resemble free n-alkanes, whereas longer chains are more conformationally ordered. It was further reported that these SAMs contain a detectable amount of gauche defects near the surface, which decreases with increasing chain lengths.

Recent studies on the selectively deuterated self-assembled silver n-octadecanethiolate layered materials also demonstrated that these materials are characterized by a lower number of gauche conformers in the alkyl chain near the solid surface.[112] Our present experimental data can be further compared with those from a recent studies of SAMs on Au nanoparticles.[113,114] In that work the CD_2 stretching modes of n-octadecanethiols deuterated at different positions [(i) positions C-2 to C-18, (ii) position C-1, and (iii) positions C-10 to C-13] bounded on the surface of Au nanoparticles were analyzed in order to monitor the thermally induced onset of local disorder in the n-alkyl chains.

The temperature dependence of the CD_2 symmetric stretching frequency of the deuterated $C_{18}S/Au$ nanoparticles confirmed that the calorimetric phase transition implicated a thermally induced change from a predominantly all-trans conformation to a chain disordered state. $C_{18}S/Au$ nanoparticles, whose chains have been deuterated only from the C_{10} to C_{13}, showed a much sharper disordering transition than particles with perdeuterated $C_{18}SH$. Moreover, the stretching frequencies observed for $1,1-d_2-C_{18}S/Au$ did not change within the temperature range from 10 - 90 °C, i.e., the conformational order is maintained at the carbon next to the gold surface. This strongly suggests that the conformational order decreases gradually with the distance from the solid support, and it is concluded that the chain disorder originates from the chain terminus region and propagates toward the middle of the chain as the temperature increases. This latter result is in contrast to the findings from our present work, where the methylene segments in the vicinity of the silica surface (C-4) exhibit a higher conformational disorder than the methylene groups in the inner part of the n-alkyl chains. It is still open whether these differences stem from different packing densities as a result of a different surface coverage (surface coverage data are not given in ref. 113) or from the different nature (i.e. bond strength) of the $C_{18}S/Au$ bond. In the latter case, the chains might be able to move laterally on the gold surface. The chains then would reposition themselves to achieve optimal interaction for a low energy state along with a higher conformational order near the surface. The situation for the present alkyl modified silica gels is different, since the $\equiv Si-O-SiX_2R$ bonds are less subject to formation/reformation.

These different conformational properties of $C_{18}S/Au$, silver n-octadecanethiolate and n-alkyl modified silica gels also might reflect the use of different solid supports. In fact, the influence of solid supports, like zirconia, titania and silica, on the conformational properties of C_{30} SAMs are examined by FT IR spectroscopy (see Chapter 5.1).[89]

In a recent molecular dynamics (MD) work on modified silica gels a characteristic variation of the amount of gauche conformers across the alkyl chains was reported.[32] This MD simulation predicted about 17% of gauche conformers at position C-4 for the C_{18} modified system at 300 K and with a similar surface coverage as used during our present work. For position C-6 a slightly larger value was reported, while the amount of gauche conformers increases to about 30% near the chain ends. Likewise, the lattice theory by Dill[75] predicted an increase of chain disorder with distance from the solid surface. Both results are different from our observations, where – independent of the actual preparation method – in general higher amounts of gauche conformers were derived, and a maximum chain disorder is found near the silica surface.

On the basis of our present FT IR investigations on n-alkyl modified silica gels we therefore assume that at low temperatures the inner parts (i.e., from positions C-6 up to C-12) of the C_{18} and C_{22} chains are characterized by a dense chain packing (i.e., chain regions with high conformational order). At the same time, the methylene segments close to the silica surface – see high conformational disorder at position C-4 (CD_2 rocking data) – are conformationally disordered even at low temperatures, which might be a prerequisite for the above-mentioned packing of the inner chain segments. Moreover, in agreement with a former extensive CH_2 wagging band analysis[23] there is also a high degree of conformational freedom near the chain ends, as expressed by the high amount of end-gauche conformers. At higher temperatures the inner chain segments gain in conformational flexibility, as can be deduced from the CD_2 stretching and rocking data. At the chain ends, i.e., beyond position C-12, the changes in conformational order with temperature are less pronounced, since here a considerable amount of conformational disorder exists (i.e., end-gauche conformers) even at low temperatures.[23]

A systematic study about the influence of pressure on the conformational properties of the tethered alkyl chains in chemically modified silica gels so far does not exist. In a recent study on monolayer-protected gold cluster molecules (MPC), different sample preparation techniques were used such as KBr pellets, dropcast films or CCl_4 solutions.[115] The corresponding FT IR spectra (symmetric and anti-symmetric CH_2 stretching bands, room temperature measurements) revealed that the conformational order of the tethered alkyl chains increases upon applying pressure during sample preparation (KBr pellets), in quite the same way as discussed for our present silica gels. Numerous studies are known which deal with pressure-induced effects on the alkyl chain conformations in the liquid phase.[116,117] Several independent studies proved that a higher sample pressure gives rise to a higher amount of gauche conformers, since liquid n-alkanes then adopt a more globular shape. This property is found to be almost independent of the actual chain length, as demonstrated by studies on n-hexane, n-pentadecane and n-hexadecane.[118-120]

For the present chemically modified silica gels we obviously have an opposite trend for the change of the alkyl chain conformation upon application of external pressure. The increase of conformational order with increasing pressure can be explained by an improved chain packing along with an increase of attractive intermolecular alkyl chain interactions and a more favorable overall molecular energy of the system. The chemical attachment of the alkyl chains on the solid support is certainly also very important for this characteristic change in alkyl conformational order upon sample pressure.

In summary, a variable FT IR study has been performed in order to examine the conformational order of alkyl chains attached to silica surfaces in the dry state as a function of chain length, chain position, and sample temperature as well as external pressure during sample preparation. Through CH_2 wagging band analysis the number of kink/gtg, double-gauche and end-gauche conformers and total number of gauche conformers per chain were derived for the octadecyl-modified sample. It was observed that the amount of kink/gtg and double-gauche conformers per chain increased upon increase of temperature and decrease upon application of pressure. The analysis of the CD_2 stretching and rocking bands provides site-specific information of the conformational order. Apart from a characteristic dependence of the conformational order from the actual chain position and temperature, the amount of gauche conformers again is considerably reduced by application of pressure. In general, it can be concluded that the systems adopt a higher ordered conformational state upon increase of the sample pressure due to an overall gain in intermolecular chain interactions. Further studies along this line, in particular addressing the impact of the sample pressure on the molecular behavior of such alkyl modified silica gels, are required in order to get a more detailed picture about the molecular features of these materials during the chromatographic separation.

5.4   Effect of Surface Coverage on C_{18} Systems

A variable temperature FT IR study of C_{18}-alkyl modified silica gels with surface coverages varying from 2 to 8.2 μmol/m^2 is presented here. Furthermore, the bonded phases are characterized by solid-state ^{13}C and ^{29}Si NMR spectroscopy. The influence of surface coverage on the conformational order of C_{18}-alkyl modified silica gels in the dry state is evaluated over the broad temperature range between 193 and 353 K. Both qualitative and quantitative studies are performed using the CH_2 symmetric and anti-symmetric stretching bands and CH_2 wagging band analysis, respectively. In addition, chromatographic shape selectivity measurements are included.[121]

^{29}Si CP/MAS NMR spectroscopy can be exploited for the determination of the surface species of modified silica gels. It provides information about the degree of cross-linking of silanes during the modification step. The ^{29}Si CP/MAS NMR spectra of five C_{18}-alkyl modified silicas recorded at room temperature are shown in Figure 5.26. Inspection of this figure reveals the presence of various types of silicon atoms which differ by their coordination sphere as a result of the silylation step. The different structural elements are illustrated in Figure 3.7. The ^{29}Si NMR signals of the trifunctional silanes (T) are located in the range from -45 to -70 ppm and refer to tri-functional groups without (T^1), with partial (T^2), or with complete cross-linking (T^3), at about -48, -55 and -67 ppm, respectively. The signals from the native silica are located in the range from -91 to -110 ppm. A gradual increase in the intensity of the T^2 and T^3 units is observed when the surface coverage increases, which indicates that the degree of cross-linking is related to the corresponding surface coverage. The highest degree of cross-linking is thus observed for the stationary phase with the highest surface coverage (8.2 μmol/m²), as reflected by intense T^3 and T^2 signals. Relatively little evidence for crosslinking is indicated for the sample with a surface coverage of 2.7 μmol/m². This stationary phase was prepared under anhydrous conditions, which would limit crosslinking. C_{18}-alkyl modified silica with 2.0 μmol/m² surface coverage was prepared with the addition of water, but with a limited quantity of the silane. Slightly increased crosslinking is indicated in the ^{29}Si NMR spectrum in Figure 5.26, even though the surface coverage is reduced compared with the 2.7 μmol/m² sample.

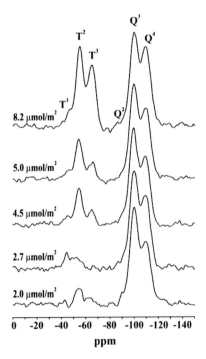

Figure 5.26: ^{29}Si CP/MAS NMR spectra of C_{18}-alkyl modified silica gels.

Differences in the alkyl chain conformational order of the C_{18}-alkyl modified silica samples can be discussed on a qualitative basis by the comparison of the ^{13}C NMR resonances from the inner methylene segments C-4 to C-15, attributed to "crystalline-like" trans and "solution-like" trans-gauche conformations.[20,22,56] A signal at about 32 ppm thus reflects alkyl chains in the trans conformation, while gauche conformations are characterized by an upfield signal at about 30 ppm (see discussion below).

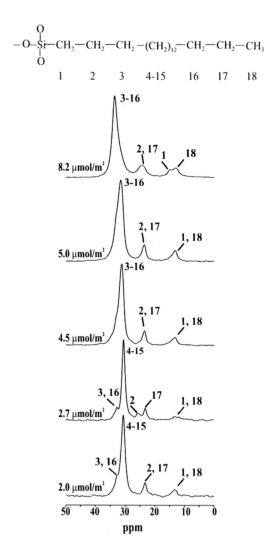

Figure 5.27: ^{13}C CP/MAS NMR spectra of C_{18}-alkyl modified silica gels.

Figure 5.27 shows the ^{13}C CP/MAS NMR spectra of the C_{18}-alkyl modified silica gels with surface coverages between 2 and 8.2 μmol/m². In addition to the dominant resonance of the inner methylene segments, a weak signal at 32.7 ppm attributed to methylene carbons C-3 and C-16 can be discerned for the 2 and 2.7 $\mu mol/m^2$ stationary phases. This signal is obscured at higher surface coverages by the ^{13}C resonance of the remaining inner methylene segments (C-4 to C-15), which subsequently shifts to lowfield with increased surface coverage, as discussed below. For the C_{18} alkyl modified silica gels with surface coverages of 2.0, 4.5, and 5 $\mu mol/m^2$, the signal at 23.5 ppm is attributed to methylene units C-2 and C-17, and the resonance at 13 ppm is due to position C-1 and the terminal methyl group C-18. The sample with 2.7 $\mu mol/m^2$ surface coverage behaves somewhat differently, as the C-2 and C-17 methylene segments show up as two separate resonances at 25.7 and 23.2 ppm. Although the reason for this behavior is not clear, similar spectra have been reported for liquid octadecylsilanes[41] which suggests a liquid like behavior for the stationary phase. The stationary phase with 2.7 $\mu mol/m^2$ was prepared under anhydrous conditions with little crosslinking, and the alkyl chains may be more randomly distributed on the silica surface than with crosslinked stationary phases. Alkyl chain distribution is thought to depend on surface silanol locations, silanol reactivity, and steric effects during synthesis. In contrast, stationary phases prepared by solution polymerization may be more organized due to self-assembly of silane polymer in solution prior to deposition and bonding to the silica surface.[4] For the 8.2 $\mu mol/m^2$ sample, the signal at 24.5 ppm characterizes the methylene units C-2 and C-17, and the separate signals at 14.8 and 12.8 ppm are due to segment C-1 and the terminal methyl group C-18.

The ^{13}C CP/MAS NMR spectra of the present phases show substantial differences for the conformational sensitive ^{13}C resonance of the inner methylene units C-4 to C-15. The variation of ^{13}C chemical shifts with surface coverage is plotted in Figure 5.28. For the lowest surface coverage of 2 $\mu mol/m^2$, the signal position is at 30.6 ppm, which indicates a high proportion of gauche conformers. This is also supported by the smaller NMR line width indicating a higher mobility of these conformationally disordered chains. Former solid-state ^{13}C NMR investigations on silica gels modified with even longer alkyl chains - C_{22}, C_{30}, and C_{34} - revealed two distinct signals for the inner methylene groups of the alkyl chain that can be assigned to crystalline-like trans and solution-like trans-gauche conformations.[20,22,56] A similar signal splitting of these methylene resonances is not observed for the present C_{18}-alkyl modified silica gels. Rather, a gradual lowfield shift of the methylene signal is observed from 30.5 ppm for 2 $\mu mol/m^2$ surface coverage to 33.5 ppm for 8.2 $\mu mol/m^2$. The latter value for the highest surface coverage indicates the presence of chains in a predominantly trans conformational state. This result is consistent with the findings for C_{22} phases for which similar lowfield shifts were observed with increased surface coverage.[22] The data for the 8.2 μmol/m^2 sample can be compared with the ^{13}C NMR data of alkanethiols [CH$_3$(CH$_2$)$_{17}$SH] attached to gold particles. Methylene carbon signals occur between 33.5 and 33.8 ppm indicates alkyl chains in a conformationally highly ordered state.[122] Likewise, crystalline n-alkanes exhibit a ^{13}C resonance for the inner methylene segments between 33 to 34 ppm.[101]

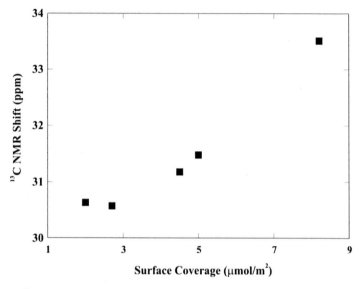

Figure 5.28: [13]C chemical shift values (ppm) plotted versus surface coverage (μmol/m^2).

The lowfield shift in the ^{13}C CP/MAS NMR spectra of the present phases can be explained through the well-known γ-gauche effects (see Chapter 3.2.2). The lowfield shift observed for the 2 μmol/m^2 sample is slightly greater than expected, based solely on surface coverage, and indicates slightly greater conformational order compared to the 2.7 μmol/m^2 sample. The solution polymerized sample (2.0 μmol/m^2) may be more ordered preassembly resulting from crosslinking during surface modification. The modified surface could consist of patches with closely spaced alkyl chains which are more ordered than randomly positioned, isolated chains.

In addition to the lowfield shift of the inner methylene groups resonance, the signal for the terminal methyl groups is split into two resonances (14.8 and 12.9 ppm) at the highest surface coverage (8.2 μmol/m^2). Albert et al.[123] discussed dual methyl resonances at 14.1 and 12.5

ppm for a monomeric C_{18} phase (surface coverage 2.4 μmol/m^2). As the signal at 14.1 ppm was found to increase at lower temperatures, it was attributed to chains in more ordered conformations. The signal at 12.5 ppm was attributed to methyl groups in a more mobile environment. This explanation was based on earlier findings for methyl groups in polyethylene, for changes from a crystalline to a non-crystalline environment.[106] The split up of methyl resonance was attributed to the heterogeneity of the sample, i.e., the presence of methyl groups with different molecular mobility due to a different chain packing. Pursch et al.[20] discussed in detail methyl group resonances of C_{30} SAMs. Dual methyl resonance profiles were observed for C_{30} SAMs on ProntoSil (silica gel) with chemical shifts of 14.7 and $\approx$ 13 ppm.

Additional effort was devoted to the C-17 resonance in order to provide more information about the conformational order of the alkyl chains. For the present C_{18}-alkyl modified silicas, the signal shifts slightly from 23.3 to 24.4 ppm with increased surface coverage, indicating the presence of chain end gauche defects even at the highest surface coverage. This is similar to the data reported for C_{18} thiol chains on a gold surface in which the C-17 resonance was observed at 24.2 ppm.[122] In the case of pure n-alkanes, the signal for C-17 is observed at 22.6 ppm in the liquid state with conformationally disordered chains and at 25.0 ppm in the crystalline state with all-trans chains.[101]

Conformational order was further examined by the position of the CH_2 stretching band in FT IR spectra. For completely disordered structures, the absorption maximum of the CH_2 anti-symmetric stretching band occurs at 2924 cm^{-1}, the value of liquid alkane, whereas for conformationally ordered systems the wavenumber is shifted to lower values $\approx$ 2915 cm^{-1} for crystalline alkanes. An analogous shift is observed for the CH_2 symmetric stretching band. Thus, the frequency shift of the band maxima of the symmetric (2854 - 2846 cm^{-1}) and anti-symmetric (2924 - 2916 cm^{-1}) CH_2 stretching bands can be used as a qualitative measure for changes in the conformational order that may result from changes in surface coverage or sample temperature.[23,89]

Figure 5.29 shows the FT IR spectra of the C_{18}-alkyl modified silica samples, recorded at room temperature, covering the CH_2 symmetric and anti-symmetric stretching band regions. The CH_2 anti-symmetric stretching band maximum for the low loaded C_{18} phases is $\approx$ 2926 cm^{-1}, whereas the band maximum for the 8.2 μmol/m^2 sample is $\approx$ 2920 cm^{-1}. CH_2

bandwidths decrease with increasing surface coverage. These findings indicate that conformational order increases with increased surface coverage.

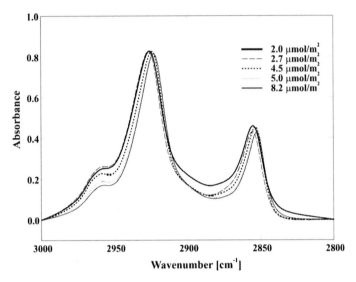

Figure 5.29: Influence of surface coverage on the symmetric and anti-symmetric CH₂ stretching modes of C₁₈-alkyl modified silica gels.

Temperature plays an important role during chromatographic separations, as shape selectivity can be influenced by column temperature, especially for certain classes of solutes with rigid and well-defined molecular shapes.[24] Therefore, the present stationary phase materials with different surface coverages were studied by variable temperature FT IR spectroscopy over the temperature interval 193 to 353 K. CH₂ anti-symmetric and symmetric stretching band frequencies for the C₁₈-alkyl modified silica gels are summarized in Figures 5.30 and 5.31. The absorption band maxima for all C₁₈ phases are found to shift towards higher wavenumbers with increasing sample temperature. For instance, the sample with 2 μmol/m^2 coverage exhibits an anti-symmetric absorption band maximum at 2922.9 cm^{-1} at 193 K, and at 2927.0 cm^{-1} at 353 K. For all samples, the overall change in the CH₂ anti-symmetric stretching band position is about 4 cm^{-1} over the temperature interval studied. The shift in the band position occurs uniformly with temperature, which is consistent with an absence of a phase transition for the present stationary phase materials. It is interesting to note that the CH₂ stretching bandwidth increases at higher temperatures as previously reported.[89] Both observations, the broadening of the absorption bands and their temperature dependent shift, demonstrate that the tethered alkyl chains become more disordered upon increase of the sample temperature. Figures 5.30 and 5.31 further illustrate that the absorption frequencies of the symmetric and anti-symmetric bands for higher surface coverage C_{18} phases are lower than those of the other C_{18} phases over the entire temperature range studied. However, based on the symmetric and anti-symmetric band frequencies, greater conformational order is apparent in the 2.0 $\mu mol/m^2$ sample than in the 2.7 $\mu mol/m^2$ sample. This finding is consistent with the ^{13}C NMR data for these samples, but is not expected based solely on surface coverage.

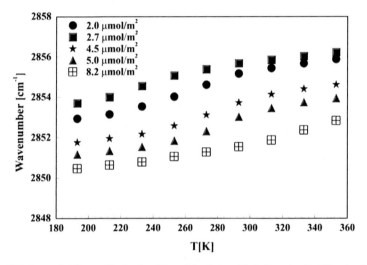

Figure 5.30: Symmetric CH_2 stretching band positions of C_{18}-alkyl modified silica gels with different surface coverages.

The present IR data can be further compared with data from a study of C_{18} thiol chains on gold particles.[122] On the basis of the CH_2 symmetric and anti-symmetric stretching band positions for the C_{18} thiol chains (2850 and 2918 cm^{-1}, respectively) and the present C_{18}-alkyl modified silica gels with the highest surface coverage (2851.5 and 2920.7 cm^{-1}, respectively), it is anticipated that the C_{18} thiol chains have higher conformational order, which is consistent with the findings from the aforementioned ^{13}C NMR data (see above).

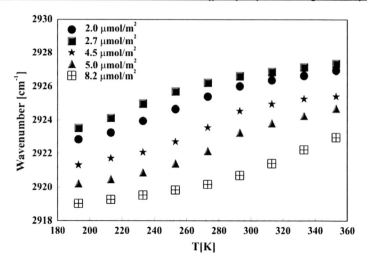

Figure 5.31: Anti-symmetric CH_2 stretching band positions of C_{18}-alkyl modified silica gels with different surface coverages.

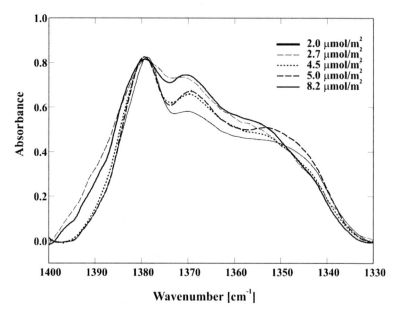

Figure 5.32: Surface coverage dependence of the experimental CH_2 wagging band spectra of C_{18}-alkyl modified silica gels for different surface coverages.

Representative experimental FT IR spectra, covering the wagging band region between 1300 and 1400 cm^{-1} and which were recorded at room temperature, are displayed in Figure 5.32.

78

Inspection of these spectra reveals that for the C_{18}-alkyl modified silica with the highest surface coverage the relative intensity of the band at 1368 cm⁻¹, characterizing the kink/gtg conformers, is lower than those for the stationary phases with lower surface coverages. A similar trend is observed for the bands at 1353 and 1341 cm⁻¹, attributed to double-gauche and end-gauche conformers, respectively. In general, these experimental FT IR spectra indicate that conformational order is correlated with surface coverage.

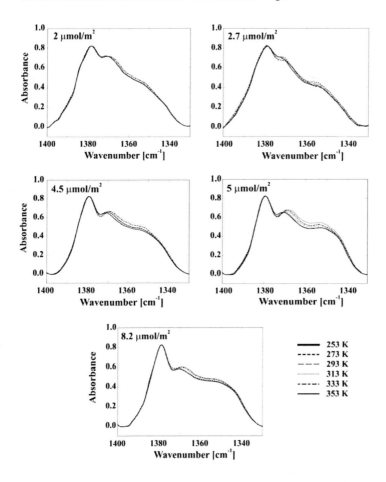

Figure 5.33: Temperature dependence of the experimental CH_2 wagging band spectra of C_{18}-alkyl modified silica gels for different surface coverages.

The influence of temperature on CH₂ wagging band intensities is shown in Figure 5.33. Wagging band intensities are observed to increase as a function of sample temperature for all samples, which is indicative of an increase in the conformational disorder of the alkyl chains.

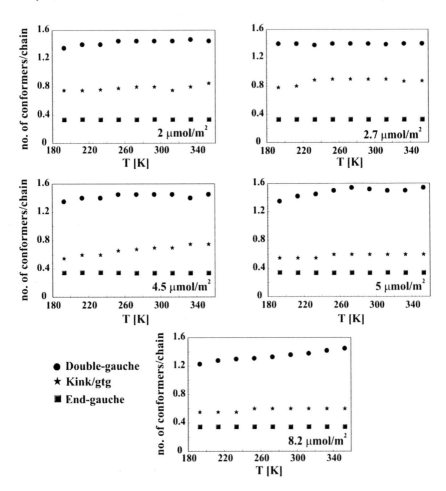

Figure 5.34: Number of kink/gtg, double-gauche, and end-gauche conformers per chain in C₁₈-alkyl modified silica gels with different surface coverages.

The results from the curve fitting analysis of the CH₂ wagging band regions are summarized in Figure 5.34. The number of end-gauche conformers per chain (about 0.35) is found to remain almost unaffected by sample temperature and surface coverage, whereas the number of kink/gtg and double-gauche conformers per chain decreases with increasing surface coverage and increases with sample temperature. For example, for the 2 μmol/m² sample, the number of kink/gtg conformers per chain varies between 0.75 (193 K) and 0.85 (353 K). For the 8.2 μmol/m² sample, the number of kink/gtg conformers per chain varies between 0.45 and 0.55 over this temperature interval. Similar variations in the number of double-gauche conformers per chain are observed with changes in temperature. Notably, the fraction of double-gauche conformers is higher than the corresponding fraction of kink/gtg conformers for all samples.

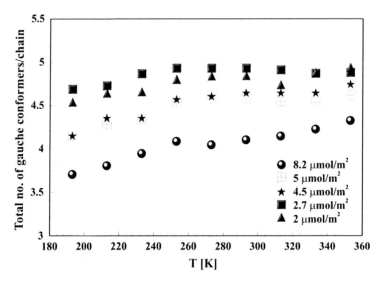

Figure 5.35: Influence of surface coverage on the total number of gauche conformers per chain for C_{18}-alkyl modified silica gels.

The total numbers of gauche conformers per chain, derived for the C_{18}-alkyl modified silicas, are depicted in Figure 5.35. As expected, the total number of gauche conformers per chain increases with lower surface coverage. For the C_{18} phases with the highest surface coverage, the values lie between 3.7 and 4.3, while for the lowest surface coverage the values are between 4.5 and 4.9. The greatest disorder is apparent for the 2.7 μmol/m² sample rather than the 2.0 μmol/m² sample, which is attributed to the greater crosslinking and thus closer alkyl chain spacing in the 2.0 μmol/m² sample.

In general, the NMR and FT IR data indicate that C_{18} phases with lower surface coverage possess a higher fraction of gauche defects, i.e., are characterized by a higher conformational disorder of the alkyl chains. This disorder arises from the larger spatial freedom and higher mobility of the alkyl chains. Conformational order increases with increased surface coverage as a consequence of constraints placed on alkyl motion by spatial restrictions and as a consequence of increased van der Waals interactions.[32]

The ^{13}C resonance of the inner methylene segments for the sample with the highest surface coverage appears at 33.5 ppm, indicating a high degree of conformational order. Alkyl chains in the all-trans conformational state are known to exhibit characteristic wagging band progressions in FT IR spectra between 1350 and 1100 cm^{-1}. Even though ^{13}C NMR data indicate that high conformational order exists for the 8.2 μmol/m^2 sample, FT IR wagging band progressions are not observed, which implies that at least a few gauche defects are present even at the highest surface coverage.[97]

The influence of alkyl conformational order on chromatographic shape recognition can be assessed through measurement of selectivity factors for shape selective solute probes. SRM 869a Column Selectivity Test Mixture for Liquid Chromatography is intended primarily for this purpose, and contains three planar and nonplanar polycyclic aromatic hydrocarbons (PAHs). The selectivity factor for tetrabenzonaphthalene (TBN) and benzo[a]pyrene (BaP; $\alpha_{TBN/BaP}$) provides a numerical descriptor that has been correlated with the ability to separate isomers and shape constrained solute mixtures (i.e., shape selectivity). Values typically range from ≈ 0.5 (high shape selectivity) to ≈ 1.7 (low shape selectivity) for most commercial C_{18} columns. Selectivity factors for columns prepared in this study are plotted as a function of surface coverage in Figure 5.36. As expected, the selectivity factors ($\alpha_{TBN/BaP}$) generally decrease with increasing surface coverage. It is interesting to note that $\alpha_{TBN/BaP}$ values for the 2.0 and 2.7 μmol/m^2 materials are consistent with the ^{13}C NMR and FT IR findings. The 2.0 μmol/m^2 column exhibits slightly greater shape recognition properties (as indicated by $\alpha_{TBN/BaP}$ values) compared with the 2.7 μmol/m^2 column, even though it has lower surface coverage. These trends are illustrated in the plots of ^{13}C NMR and FT IR band shifts as a function of $\alpha_{TBN/BaP}$ (see Figure 5.36). These data provide strong evidence for the relationship between chromatographic performance (i.e., shape recognition) and stationary phase conformational order. Thus, interactions between the solute and the stationary phase appear to be more sensitive to conformational order than in the way in which the order is achieved.

The existence of a higher conformational order (from ^{13}C NMR and FT IR data) found for C_{18} thiol chains on gold particles[122] as compared with the present C_{18}-alkyl modified silica gels most probably can be explained by the different surface characteristics of gold, different surface coverages or from the different nature of the sulphur-gold bond.

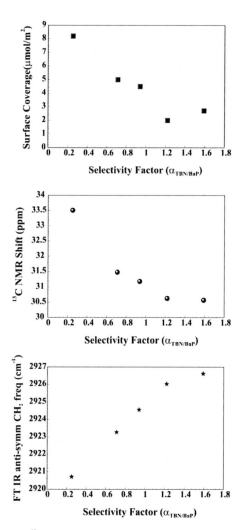

Figure 5.36: Surface coverage, ^{13}C chemical shift and FT IR anti-symmetric CH_2 frequencies plotted as a function of the shape selectivity factor $\alpha_{TBN/BaP}$ for C_{18}-alkyl modified silica gels

The partition model theory predicts that the chain segments nearest to the silica surface are highly ordered and are aligned perpendicular to the surface.[32] In addition, it is anticipated that the increase of conformational order upon increasing surface coverage covers the entire alkyl chain. This partition model theory is in qualitative agreement with the ^{2}H NMR studies on dimethyloctadecylsilyl modified silica gels.[44] It was concluded that for low loaded silica – due the van der Waals interactions between the n-alkyl chains and the silica surface - the disordered C_{18} chains interact strongly with the silica surface, giving rise to unusual motional constraints. At higher surface coverage the chains motion become less restricted, since the C_{18} chains are farther displaced from the surface. At the same time, a high chain ordering for the inner part of the alkyl chain (positions C-9, C-10) was observed. Further detailed information about the mobility and conformational order of C_9, C_{18} and C_{22} chains on silica were provided by a comprehensive variable temperature FT IR and ^{2}H NMR study using selectively deuterated and nondeuterated alkyl chains of different lengths (from C_8 to C_{30}).[21,23] In that work silica surface coverage was about a factor of two higher the high surface coverage samples from Zeigler et al.[44] It was shown that the chain conformation exhibits a distinct gradient over the alkyl chain and also depends on the actual chain length and temperature. An unusually high amount of gauche conformers was found for position C-4 close to the silica surface, which holds for all alkyl chain lengths. The inner chain segments for the longer C_{18} and C_{22} chains, as reflected by positions C-9 and C-12, again are highly ordered, while the free chain ends are characterized by a substantial amount of chain disorder. Both the high chain ordering of the inner chain segments and the substantial gauche amount at position C-4 can be understood by the van der Waals interactions between the alkyl chains. The gauche conformers at position C-4 indicate a bending of the chains which results in a more efficient alkyl chain packing on the silica surface.

In summary, solid-state ^{13}C and ^{29}Si NMR measurements as well as variable temperature FT IR studies were performed on C_{18}-alkyl modified silica gels with different surface coverages. The influence of surface coverage on the degree of cross-linking was examined by ^{29}Si NMR spectroscopy. The highest degree of cross-linking among the present samples resulted for C_{18}-alkyl modified silica gels with the highest surface coverage of 8.2 μmol/m^2. The influence of surface coverage on the alkyl chain conformational order was studied on a qualitative basis by ^{13}C NMR spectroscopy (chemical shifts of inner methylene segments) and the CH$_2$ symmetric and anti-symmetric stretching bands (band position and band width) in FT IR spectra of the tethered alkyl chains. Both methods provided consistent results, i.e., the conformational order decreases with increasing sample temperature and increases with higher surface coverage. The integral amount of kink/gtg, double-gauche, end-gauche and total number of gauche conformers over the whole alkyl chains were derived from the analysis of the wagging band spectra. For the highest surface coverage, gauche defects were still observed and wagging band progressions were absent.

5.5   Influence of Synthetic Routes

In this chapter, the influence of the synthetic route on the conformational order of C_{18} and C_{30}-alkyl modified silica gels is presented. Here, two different synthetic routes are used for synthesizing C_{18} and C_{30}-alkyl modified silica gels, namely, the so-called solution and surface polymerization routes. During the course of these investigations, the following samples were available, (i) C_{18} solution polymerized (hereafter denoted as C_{18} Solution in Figures and Tables), (ii) C_{18} surface polymerized (C_{18} Surface), (iii) C_{30} solution polymerized (C_{30} Solution), and (iv) C_{30} surface polymerized (C_{30} Surface). IR conformational sensitive bands are employed to explore the conformational order of the tethered alkyl chains. In addition, ^{13}C CP/MAS NMR measurements are included in order to get complementary information about the molecular properties of the alkyl chains. The results derived from these measurements are discussed in the context of the synthetic routes and alkyl chain lengths.

The ^{29}Si CP/MAS NMR spectra of C_{18} and C_{30}-alkyl modified silica gels (Figure 5.37) show the presence of various types of silicon atoms as a result of the silylation step. It provides information about the degree of cross-linking of the silane in the modification step, and is also applicable for determining the surface species of the modified silica gels. The intensities of the two peaks at -67 and -56.6 ppm can be used to derive the amount of silicon atoms that bear n-alkyl groups. The relative intensities of the T^2 and T^3 units are identical for both C_{18} phases, which prove that the degree of cross-linking is the same, irrespective of the synthetic routes. However, in the case of the C_{30} phases, an increase in the intensity of T^3 units is observed for surface polymerized phases when compared with the solution polymerized phases, pointing to a higher degree of cross-linking of the former ones. The relative intensity of the Q^3 units (-101 ppm) for C_{30} and C_{18} phases indicates that the surface polymerized phases exhibit a lower amount of silanol groups (low intensity signals) as compared to the solution polymerized phases.

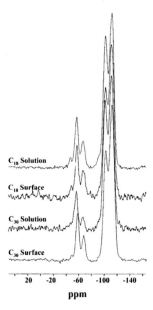

Figure 5.37: Experimental ^{29}Si CP/MAS NMR spectra of C_{18} and C_{30}-alkyl modified silica gels.

The available results from literature about the relative intensity of the Q^3 units in alkyl modified silica gels are not consistent.[17,56] For instance, Pursch et al.[17] prepared C_{18} phases by different synthetic routes, and found that the Q^3 signals for surface polymerized phases exhibit peaks of lower intensity as compared to solution polymerized phases. At the same time, solution polymerized C_{18} and C_{22} and surface polymerized C_{30} and C_{34} phases were reported to behave differently.[56] Here, Q^3 signals of low intensity were observed for solution polymerized C_{18} and C_{22} phases. Likewise, comparatively higher fractions of silanol groups were observed for surface polymerized C_{30} and C_{34} phases. It was concluded that such differences might be the result of different reaction temperatures for surface polymerization reactions with C_{18} and C_{22} silanes (ambient) and C_{30} or C_{34} silanes (gentle reflux).

Variable temperature ^{13}C CP/MAS NMR spectra of the C_{18} phases (solution and surface polymerized) are shown in Figure 5.38. The NMR signal for the main $(CH_2)_n$ group for the C_{18} surface polymerized phases at room temperature occurs at 32.6 ppm corresponding to crystalline-like trans conformations. The signal observed at about 23.8 ppm characterizes the methylene units C-2 and C-17 and the signal at 12.8 ppm shows the presence of the C-1 group and the terminal methyl group. The main resonance frequency of the $(CH_2)_n$ groups is shifted upfield gradually upon increase of temperature and reaches 31.6 ppm at 350 K. This signal indicates higher alkyl chain mobility and the presence of a large fraction of trans-gauche conformations when compared to the signal observed at 32.6 ppm for room temperature. In addition, the signal of methylene groups C-2 and C-17 observed at 23.8 ppm for room temperature varies with the sample temperature and reaches 23.2 ppm at 350 K. These observations lead to the conclusion that chain disorder occurs with increase in temperature in all segments of the chain, irrespective of the methylene group location.

In the case of C_{18} solution polymerized phases, the main $(CH_2)_n$ resonance appears at 31.4 ppm corresponding to trans-gauche conformations. Comparison of the two synthetic routes reveals that the C_{18} surface polymerized phases possess greater conformational order than C_{18} solution polymerized phases. Figure 5.38 further demonstrates that the amount of gauche conformations again increases with increasing temperature, since the main resonance frequency shifts slightly upfield with a value of 31.0 ppm at 340 K.

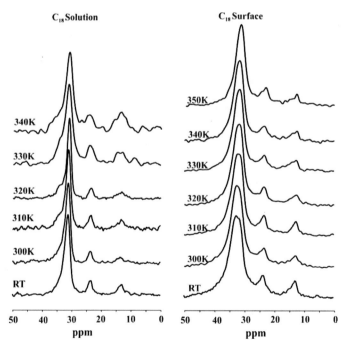

Figure 5.38: Variable temperature ^{13}C CP/MAS NMR spectra of the C_{18} solution and surface polymerized phases.

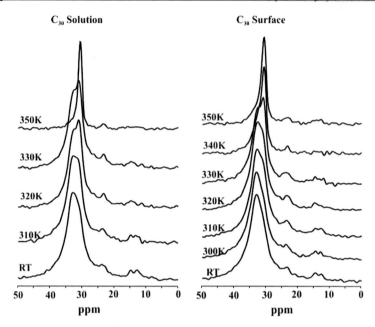

Figure 5.39: Variable temperature ^{13}C CP/MAS NMR spectra of the C_{30} solution and surface polymerized phases.

The main $(CH_2)_n$ resonance for the C_{30} surface polymerized phases (33.1 ppm) and the C_{30} solution polymerized phases (33.0) are practically identical and obviously do not depend on the synthetic route (see Figure 5.39). The main resonance frequency for both C_{30} phases gradually decreases to 30.8 ppm with increasing temperature, pointing to the presence of gauche conformations. The existence of both conformers, i.e., trans and gauche, are clearly visible in both C_{30} phases especially at 330 K.

NMR signals for the terminal methyl group are observed at 14.7 and 12.9 ppm, in contrast to a single resonance at 12.8 ppm for the C_{18} solution and surface polymerized phases. Such a split-up of the methyl signals was also observed in earlier studies of C_{18}-alkyl modified silica gels especially at the highest surface coverage of 8.2 µmol/m^2 (see Chapter 5.4).[121] The occurrence of dual methyl resonance was studied in detail for monomeric C_{18} phase[123] (surface coverage: 2.4 µmol/m^2) and for C_{30} SAMs[20] as an important characterization tool to understand the behavior of alkyl chain structure. Based on the earlier findings, the dual methyl resonance has been attributed to the heterogeneity of the sample, i.e., methyl groups of different mobilities exist in these materials.

The comparison of the main $(CH_2)_n$ resonance of C_{18} and C_{30} phases for the two solution polymerized phases, reveals a lowfield shift of 1.8 ppm for the C_{30} phases. This can be explained by the influence of alkyl chain lengths on the conformational order of alkyl modified silica gels. In addition, upfield shifts are observed on increasing the temperature for all samples. Changes in alkyl chain conformational order that results from change in alkyl chain length and temperature can be evaluated in terms of γ-gauche effects (see Chapter 3.2.2). Also, a smaller NMR line width is observed for C_{18} solution polymerized phases indicating a higher mobility of the conformationally disordered chains.

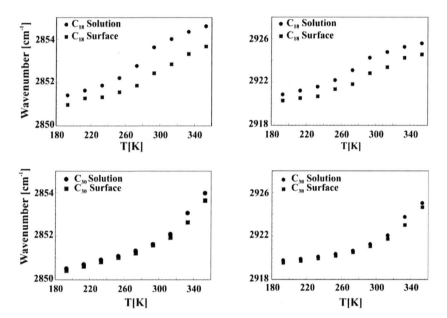

Figure 5.40: Symmetric and anti-symmetric CH_2 stretching band positions of C_{18} and C_{30} solution and surface polymerized phases.

Symmetric and anti-symmetric CH_2 stretching bands are utilized for the study of the influence of alkyl chain lengths, synthetic routes and sample temperature. The conformational order in the C_{18} and C_{30} phases are assessed from the position of the CH_2 stretching bands. The absorption band maxima of the anti-symmetric CH_2 stretching bands for C_{18} phases at room temperature are 2924.3 and 2922.8 cm^{-1} for solution and surface polymerized phases, respectively. As mentioned earlier (see Chapter 3.1.1), the shift of the band maximum to lower wavenumbers for the surface polymerized phases is indicative of lesser gauche contents, reflecting a higher conformational order of the alkyl chains at this sample temperature.

The frequency of the CH_2 anti-symmetric stretching for C_{30} phases at room temperature is 2921.1 and 2921.3 cm^{-1} for solution and surface polymerized phases, respectively. Again, it can be concluded that the synthetic routes have no influence on the conformational order of the C_{30} alkyl chains.

Furthermore, conformational order of C_{18} and C_{30} phases are examined over a broad temperature range. The absorption frequency increases with increasing temperature, irrespective of the synthetic routes and alkyl chain lengths for the whole temperature range. The wavenumbers of the symmetric and anti-symmetric bands of the C_{18} surface polymerized phases are lower than that of C_{18} solution polymerized phases (see Figure 5.40). In the case of the C_{30} phases, the symmetric and anti-symmetric CH_2 stretching bands are nearly similar for both solution and surface polymerized phases, i.e., the conformational order is the same irrespective of the synthetic routes. In addition, the anti-symmetric CH_2 stretching bands of the C_{30} phases are found at lower wavenumbers than those of the C_{18} phases. From these observations, it is concluded that the alkyl chain length strongly affects the conformational order and mobility of tethered alkyl chains.

A curve fitting analysis of the CH_2 wagging band regions was performed in order to get the amount of gauche conformers in the C_{18} and C_{30} phases. The results obtained from the analyses are summarized in Figure 5.41. The amount of kink/gtg and double-gauche conformers per chain varies strongly with the synthetic route and alkyl chain length. In contrast, the amount of end-gauche conformers per chain (about 0.33) is found to be unaffected by the actual sample temperature, alkyl chain length and synthetic route. The number of kink/gtg conformers varies from 0.54 to 0.6 at temperatures between 193 and 353 K for C_{18} surface polymerized phases, whereas the amount of double-gauche conformers remains approximately at 1.4 over the entire temperature range. Higher amounts of kink/gtg conformers are obtained in the case of the C_{18} solution polymerized phases, for which the registered values are between 0.74 and 0.83. In general, a higher amount of gauche conformers exists in the solution polymerized phases than in the surface polymerized phases.

The temperature dependence of the kink/gtg conformers in the C_{30} solution and surface polymerized phases is less pronounced, with values between 0.55 at 193 K and 0.7 at 353 K.

The amount of double-gauche conformers for both C_{30} phases is found to be the same. It is further registered that the number of double-gauche conformers is higher than the corresponding kink/gtg conformers, which holds for C_{30} both phases. Upon further comparison between the C_{18} and C_{30} phases, it can be registered that the C_{18} phases generally possess a higher amount of gauche conformers than the C_{30} phases, irrespective of the synthetic routes.

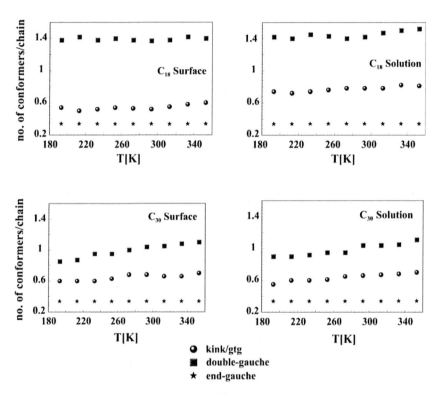

Figure 5.41: Number of kink/gtg, double-gauche and end-gauche conformers per chain in C_{18} and C_{30} solution and surface polymerized phases.

The total numbers of gauche conformers per chain for the C_{18} and C_{30} solution and surface polymerized phases are shown in Figure 5.42. Among the C_{18} phases, the surface polymerized phases have a lower amount of total number of gauche conformers, with values between 4.1 and 4.4 gauche conformers per chain. The highest alkyl chain disorder is observed for C_{18} solution polymerized phases with values between 4.65 and 5.0 gauche conformers per chain.

For C_{30} solution and surface polymerized phases, the observed total numbers of gauche conformers remains the same over the whole temperature range covered here.

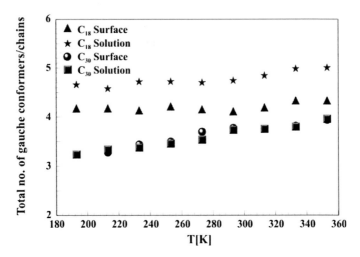

Figure 5.42: Variation of total number of gauche conformers as a function of alkyl chain lengths and synthetic routes.

Table 5.1: Properties of stationary phase materials which are used in the present discussions.

C_{18} and C_{30} phases[a]		C_{18} phases[b]		C_{18} phases[c]	
Synthetic routes	Surface coverage ($\mu mol/m^2$)	Synthetic routes	Surface Coverage ($\mu mol/m^2$)	Sample	Surface coverage ($\mu mol/m^2$)
C_{18} Surface	5.21	Solution	2.0	Surface	6.45
C_{18} Solution	4.22	Anhydrous	2.7	Solution	5.26
C_{30} Surface	4.07	-	4.5	Surface	5.00
C_{30} Solution	3.56	Solution	5.0	Solution	4.17
		Surface	8.2		

a) Present work
b) See Chapter 5.4
c) See references 17 and 28

Based on the present investigations, it is not possible to evaluate the influence of synthetic routes on the morphology of alkyl modified phases in a more precise way due to the lack of C_{18} surface and solution polymerized phases with similar surface coverage. Nevertheless, it should be noted that the synthetic route has an apparent influence on the phases, as reflected by the fact that a higher surface coverage can be only achieved with the surface polymerization route. Fortunately, a series of C_{18} phases with different surface coverages are available from our previous studies (see Chapter 5.4).[121] The influence of synthetic routes on the morphology of alkyl modified stationary phases can be evaluated by comparing the earlier data (see Chapter 5.4)[121] and other studies.[17,28] A ^{13}C resonance at 31.5 ppm was observed for C_{18} solution polymerized phases with surface coverage of 5.0 μmol/m^2 (see Chapter 5.4), which is about 1 ppm less than the signal (32.6 ppm) that observed for the present C_{18} surface polymerized phases (surface coverage: 5.2 μmol/m^2). This means that even though the surface coverage is similar, drastic difference in $(CH_2)_n$ resonances is noticed, which could be explained through the influence of synthetic route on the alkyl chain conformational order. As mentioned earlier in Chapter 2, in solution polymerization deposition of the silane polymer on the silica surface would result in a surface with some heterogeneity. In the case of surface polymerization, since the water molecules forms a monolayer coverage on the silica surface, a more regular bonded surface and a higher surface coverage would result. A homogeneity in the bonded surface from this surface polymerization technique could be the reason for the observed higher alkyl chain conformational order in C_{18} phases.

Similar results were reported earlier from other studies on solution and surface polymerized C_{18} phases.[17] It was found that two dissimilar C_{18} phases, i.e., one with trichlorooctadecyl silane by solution polymerization and another sample with dimethyloctadecylchlorosilane by surface polymerization, exhibit significant chromatographic differences despite the similar surface coverage. A lowfield shift of 0.3 ppm for the $(CH_2)_n$ resonance was observed for C_{18} solution polymerized phases as compared to surface polymerized phases. Moreover, ^{1}H NMR data for these two phases are in agreement with the ^{13}C NMR data. It was concluded that differences in stationary phase homogeneity occur with solution and surface polymerization synthetic routes.

The present results are in complete agreement with the earlier results. However, the differences observed for the ^{13}C NMR $(CH_2)_n$ resonances are larger in our case (1 ppm) when compared to the earlier data (0.3 ppm). This observation proves that surface coverage is not the only factor that governs the conformational order of alkyl modified silica gels at a given alkyl chain lengths and temperature. Other factors like synthetic route also play an important role for the conformational order and mobility of alkyl chains.

Ducey et al.[28] performed Raman measurements for the same set of samples for studying the effect of temperature, surface coverage and preparation procedure. Only slight differences were noticed by conformational indicators such as CH_2 stretching frequency and the intensity ratio $I[v_a(CH_2)]/I[v_s(CH_2)]$ among the two interesting C_{18} phases. It was concluded that conformational order and interchain coupling of the alkyl component of the C_{18} phases are sensitive to temperature and surface coverage. In addition, they concluded that the preparation procedure and nature of the alkylsilane precursor did not significantly affect the chain conformational order or interchain interactions.

In the following, the C_{30} surface polymerized phases (4.1 $\mu mol/m^2$) from the present study are compared with data from the C_{18} surface polymerized phases with the surface coverage of 8.2 $\mu mol/m^2$ (see Chapter 5.4).[121] The aim of this comparison is to elucidate the dominant factor (either surface coverage or alkyl chain length) for the conformational order of stationary phases. The $(CH_2)_n$ resonance for C_{30} surface polymerized phases is observed at 33.1 ppm. For C_{18} phases (8.2 $\mu mol/m^2$), the $(CH_2)_n$ resonance is observed at 33.5 ppm, which implies that the C_{18}-alkyl modified silica gels exhibit a higher conformational order as compared to C_{30} phases. Another comparison of C_{18} phases is possible with C_{34} surface polymerized phases (coverage: 3.5 $\mu mol/m^2$),[56] where the $(CH_2)_n$ resonance was observed at 32.6 ppm with a small shoulder at 30 ppm. On the basis of these results, it can be concluded that the influence of the surface coverage dominates over the influence of alkyl chain lengths.

Figure 5.43 illustrates that the absorption frequencies of the anti-symmetric bands for C_{18} surface polymerized phases (8.2 $\mu mol/m^2$) are lower than the C_{30} surface polymerized phases (4.1 $\mu mol/m^2$) over the entire temperature studied. This finding is consistent with the ^{13}C NMR data of these samples (see above). However, it is found that the total number of gauche conformers for C_{18} surface polymerized phases, calculated from the curve fitting analysis of the CH_2 wagging band regions possess more gauche conformers, with values between 3.7 and 4.3, than the C_{30} phases with values between 3.3 and 4.0. The differences in the number of gauche conformers observed for the C_{18} and C_{30} phases are found to be less. From the present results, it can be noted that conformational order of these C_{18} and C_{30} phases are more or less in the same range irrespective of the alkyl chain length. So far, it was believed that on increase of alkyl chain length, increase in the conformational order would result. Based on these present results, it is concluded that the surface coverage plays a major role in determining the conformational order and one should give much importance to the surface coverage when comparing the phases of different alkyl chain length.

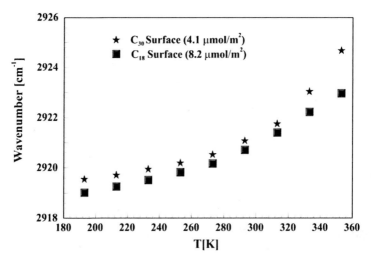

Figure 5.43: Anti-symmetric CH_2 stretching band positions of C_{18} (8.2 μmol/m^2) and C_{30} (4.1 μmol/m^2) surface polymerized phases.

In close agreement with the present results, higher alkyl chain conformational order was observed for the C_{18} phases by Cheng et al.[124] with the trans signal at 33.4 ppm and gauche signal at 30.0 ppm. It further supports our findings that influence of surface coverage dominates over the influence of alkyl chain lengths on the conformational order of stationary phase materials.

In summary, in this part the influence of synthetic routes, alkyl chain lengths and temperature on the conformational order of the tethered alkyl chains was demonstrated. As indicated by changes in the spectral properties, both from NMR and FT IR spectroscopy, the morphology of the tethered alkyl chains varies with the synthetic routes. Despite the similar surface coverage, an increase in the conformational order of the alkyl modified silica gels were observed for the C_{18} surface polymerized phases, indicating that the alkyl chains are in the more ordered conformational state with fewer gauche content and more trans bonds. The possibility of forming different conformational states reduces with increasing alkyl chain length and decreasing temperature. In addition, our investigations prove that surface coverage plays a major role on the conformational order of alkyl modified silica gels. It is found that the influence of surface coverage dominates over the influence of alkyl chain length on the conformational order of stationary phase materials.

5.6 Impact of Solvents

The influence of solvents on the conformational order of C_{18}-alkyl modified silica gels is studied by means of FT IR spectroscopy. Here, various solvents such as acetone, chloroform, cyclohexane and dichloromethane are used. In addition, perdeuterated solvents (chloroform, n-hexane, DMF, cyclohexane) are employed in order to study the deuterium isotope effect on alkyl chain conformational order of stationary phase materials. The additional advantage of using perdeuterated solvents is that overlap of some absorption bands of solvents and alkyl modified phases can be avoided since perdeuterated solvent peaks are shifted to lower energies, where they no longer interfere.

Table 5.2: Solvatochromic parameters π^*, α, β for solvents.

Solvent	π^*	α	β
Dimethylformamide	0.88	0	0.69
Dichloromethane	0.82	0	0
Acetone	0.71	0.08	0.48
Chloroform	0.58	0.44	0
Cyclohexane	0	0	0
n-hexane	-0.08	0	0

The chosen solvents exhibit different properties, such as shape, size, polarity, dipole moment, polarizability, ability for bond formation and hydrophobicity, which are classified according to solvatochromic parameters π^*, α and β, developed by Kamlet et al.[125] The parameter π^* is an index for solvent dipolarity/polarizability that measures the capability of the solvent to stabilize a charge or a dipole by virtue of its dielectric effect. The α scale describes the ability of the solvent to donate a proton to a solute molecule, whereas the β scale refers to the ability of the solvent molecules to accept a proton from the solute.

As explained in the previous sections, the conformational order of alkyl chains can be obtained from the absorption band maxima of symmetric and anti-symmetric CH_2 stretching band regions. Order and disorder of the alkyl chains can be identified with reference to the absorption band maxima of n-alkanes.[80] The absorption frequencies of the symmetric and anti-symmetric CH_2 stretching band regions for C_{18} phases with various solvents are examined at 293 K.

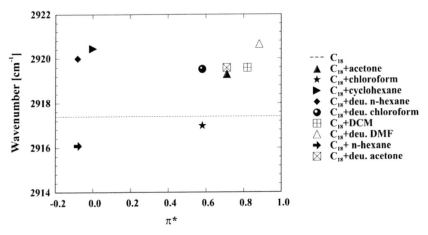

Figure 5.44: Anti-symmetric CH_2 stretching band frequencies as a function of solvatochromic parameter π^* for C_{18} phases in various solvents (T= 293 K)

The influence of the solvent on the conformational order of alkyl modified silica gels is examined on the basis of reference measurements on dry C_{18} samples under the same experimental conditions, i.e., samples were measured by using two KBr plates. The experimental anti-symmetric CH_2 stretching band positions, measured at 293 K, are plotted against solvatochromic parameter π^* to examine the correlation of the conformational order/disorder of the stationary phase materials and the solvent properties (see Figure 5.44). A very strong shift of the corresponding band maxima is observed for the sample with deuterated DMF, which has the highest π^* value of 0.88. Dichloromethane which has a slightly lesser π^* value than deuterated DMF, has a lower conformational order when compared to deuterated DMF. An increase in the conformational order with decreasing π^* is observed for acetone, chloroform and n-hexane. However, cyclohexane and perdeuterated n-hexane do not follow the trend. Even though cyclohexane and perdeuterated n-hexane have π^* values of 0 and -0.8, the absorption maxima are shifted to higher wavenumbers as we can see from this figure. The absorption maxima are found to be the same for both perdeuterated and non-deuterated acetone. For C_{18} phases in perdeuterated n-hexane or chloroform, the absorption maxima are shifted to higher wavenumbers when compared to the non deuterated n-hexane or chloroform. The same comparison could not be made for deuterated and non deuterated DMF due to the overlap of non deuterated DMF peaks with the stretching regions of the stationary phase materials.

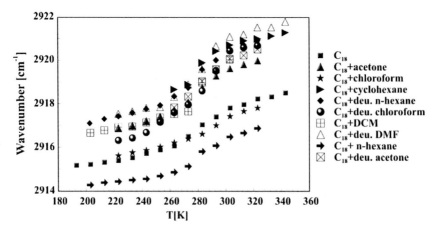

Figure 5.45: Anti-symmetric CH_2 stretching band frequencies as a function of solvents at different temperatures.

The conformational order of the chains varies with the sample temperature and is thus expected to affect the retention process in RP-HPLC. Therefore, the influence of solvent on C_{18} phases is examined over a wide temperature interval (see Figures 5.45 and 5.46). It can be seen that this temperature effect is also visible for the present samples. Moreover, the actual temperature dependence varies with the particular solvent. The influence of temperature varies on addition of solvent to the C_{18} phases. For instance, the C_{18} phases with deuterated DMF exhibits an anti-symmetric absorption maximum at 2917.5 at 223 K, and 2921.8 at 343 K. The overall change in the CH_2 anti-symmetric stretching band position is 4.3 cm^{-1} over the temperature interval studied. For dry C_{18} phases, the overall change in the CH_2 anti-symmetric stretching band position is 3.1 cm^{-1} over the same temperature interval. In addition, the influence of temperature varies with solvent. The conformational order for dry C_{18} phases and in presence of chloroform is found to be the same for temperature below 273 K, whereas at higher temperature the C_{18}/chloroform phases are higher ordered than the dry C_{18} phases.

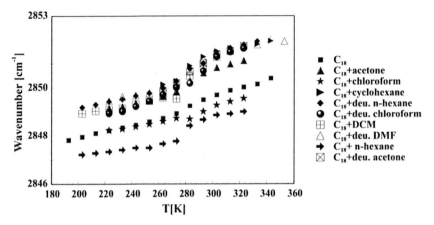

Figure 5.46: Symmetric CH_2 stretching band frequencies as a function of solvents at different temperatures.

Similar shifts are observed for C_{18} phases with various solvents in the symmetric stretching regions. In general, the trend observed from symmetric and anti-symmetric CH_2 stretching band regions are found to be the same.

A distinct effect of perdeuterated solvents as compared to non deuterated samples on the conformational order of C_{18} phases is expected on the basis of the former chromatographic studies on deuterated solvents.[126,127] Recently, a racemic mixture of the R and S isomers phenyl (phenyl-d_5) methanol was separated by HPLC using cellulose based stationary phase material. It was demonstrated that the interactions are different for the two enantiomers.[126] Baweja[127] has used RP-HPLC by employing C_{18} phases for the separation of eight isotopomer pairs of aromatic compounds. Deuterated compounds were eluted more rapidly than the non-deuterated aromatic hydrocarbons. It was speculated that the van der Waals forces were responsible for the order of elution and noted that the chromatographic behavior depends on the interaction between the C-H and C-D bonds and the stationary phases. It was stated that the oscillation frequency of C-H or C-D bond creates an electromagnetic field, which produces an induced field of opposite charge in the surrounding molecules. Since the C-H bond has a higher oscillation frequency (3300 cm^{-1}), it induces greater forces of attraction between itself and the stationary phase than with a C-D bond (2334 cm^{-1}). Due to this weaker attraction, deuterated compounds eluted earlier than the non-deuterated compound. In the present case, all perdeuterated solvents except for acetone make alkyl chains conformationally more disordered. The observed trend of deuterated solvents on the conformational order could be due to the deuterium isotopic effects, which influence the changes in interactions upon changes in the vibrational frequencies.

From the present results, it is quite obvious that the alkyl chain conformations vary on addition of the solvent to alkyl modified silica gels. For solvents such as acetone, cyclohexane, dichloromethane and deuterated solvents, the alkyl chain flexibility increases and becomes more disordered. In the case of n-hexane, it becomes more ordered. It should be noted that the observed conformational order/disorder is for the solution polymerized C_{18} phases with surface coverage of 4.7 $\mu mol/m^2$. In this context, it is worthwhile to note that the influence of solvents on the conformational order also might vary with the alkyl chain length, synthesis procedure, temperature, pressure and stationary phase support, as outlined in the previous chapters.

In the present case, an increase in the conformational order is observed for C_{18} phases in presence of n-hexane. The current observation is in accordance with studies from Lochmüller et al.,[128] in which they studied excimer decay profiles for (3-(3-pyrenyl)propyl)dimethylmono- chlorosilane [3PPS] and C_{18} modified silica in presence of various solvents. It was concluded that a collapsed state exists for C_{18} phases in presence of water, acetonitrile and methanol, whereas a more extended state exists for n-hexane and tetrahydofuran. However, Gilpin and Galgonda[129] investigated the influence of various solvents on specifically deuterated dodecyl modified silica by ^2H NMR spectroscopy and found that chains were in the most conformational disordered state in presence of n-hexane.

It is noticed from the aforementioned comparison that it is very complicate to compare the present results with recent solvent studies on alkyl modified systems due to the different parameters which play an important role on deciding the interaction of solvents with stationary phase materials. That is, it is very likely that differences in the properties of the solvents, isotope effects, surface coverage and synthesis procedure also play an important role on the influence of solvents on the conformational order of the stationary phases (see also discussion below).

Although in the present case methanol was not examined, we nevertheless briefly review the existing data about this solvent. It again demonstrates the difficulties during the data interpretation of such solvent studies. Small angle neutron scattering was used to determine the phase thickness (Å) and the volume fraction of alkyl chains within the bonded phase layer.[130] It turned out that in the presence of methanol, most of the alkyl chains are bent or disordered and thus the thickness of alkyl phases was found to be 25-35% less than that indicated by an all-trans alkyl conformation. In quite contrast, from an independent FT IR study an increase in the conformational order for C_{18} phases was derived upon addition of methanol.[24]

Ducey et al.[68] employed Raman spectroscopy to investigate the influence of a perdeuterated mobile-phase and self-associating solvents on alkyl chain rotational and conformational order in a series of C_{18} phases with different surface coverage. It was found that conformational order is dependent on solvent parameters (polarity, size, etc,), temperature, surface coverage and polymerization method. A slight increase in the conformational order was observed for C_{18} phases (1.6 $\mu mol/m^2$) in methanol and acetonitrile.[27] With nonpolar solvents tetrahydrofuran, benzene, toluene, n-hexane and chloroform, an increase in the conformational disorder was observed. In contrast, for C_{18} surface polymerized phases (6.45 $\mu mol/m^2$), alkyl chains were found to be more ordered with hexane, benzene, THF and acetone.[68] Conformational disorder was observed for all non polar solvents for C_{18} phases with the surface coverage of 5 $\mu mol/m^2$. In the case of polar solvents also irregularities on the conformational order were observed. For two C_{18} surface polymerized phases with a surface coverage of 6.45 and 5 $\mu mol/m^2$, conformational order increases in polar solvents. However, for phases prepared by solution polymerization with a surface coverage of 5.26 and 4.17 $\mu mol/m^2$, a decrease in the conformational order was observed.

Based on the present FT IR stretching band data, one cannot be sure whether partitioning or adsorption processes occur on addition of solvents to alkyl modified silica gels. Solvents might enter into a cavity (free volume) between the chains of the stationary phase causing conformational disorder in the alkyl chains or solvents might be adsorbed on the top of the surface of the alkyl chains bonded on the silica surface. Recent studies on the basis of theory of excess adsorption showed that a significant amount of organic solvents like acetonitrile, tetrahydrofuran and methanol accumulated on the top of the collapsed bonded layer. In addition, it shows that the accumulated amount was independent of the length of the alkyl chains tethered to the silica surface.[131]

In contrast Martire et al.[67] considered the alkyl modified surface as a breathing surface in which alkyl chains could become swollen by solvent penetration, and are therefore moved away from the support surface (trans conformations) in presence of non polar solvents. In the presence of polar solvents, however, alkyl chains tend to promote collapse of the chains upon each other and toward the solid supports, in order to maintain a relatively non polar character. Dill[75] has predicted theoretically and Sentell et al.[132] have shown experimentally that the degree of bonded phase alkyl chain interaction is a function of surface coverage, and that the degree of chain interaction strongly influences the ability of any solute or solvent components to penetrate the chain structure. At higher surface coverage (greater than 3.1 $\mu mol/m^2$), a decrease in the partition coefficient was observed with increasing surface coverage. It was concluded that the partitioning process was not favorable for the alkyl chains with the higher conformational order.

The degree of interaction between the solvent and the RP-HPLC materials are elusive, even though a vast number of publications are available. An attempt to study the solvent-stationary phase interaction has been made. From the present studies, it is found that the C_{18} phases and mobile phase interact significantly, resulting in changes of the conformational order of the attached alkyl chains. It is further registered that some deuterated solvents exhibit isotope effects, which are thought to arise from changes of the non-bonded interactions upon changes in the vibrational frequencies. However, a clear trend of conformational order with respect to solvatochromic parameter π^*, deuterium isotope effect is not observed. From temperature dependent studies, it is found that the conformational order also varies significantly upon solvent addition, which is more pronounced at elevated temperatures.

Chapter 6

Summary

Reversed phase liquid chromatography (RP-HPLC) is a powerful separation technique for the analysis and purification of chemical compounds. In alkyl modified stationary phase materials, the conformational order of the alkyl chain moieties is expected to be a major source for the efficiency and selectivity during chromatographic separations. Recent studies have revealed that the conformation order, chain dynamics and their concomitant effects on retention and selectivity depend on various factors such as alkyl chain length, surface coverage and temperature. Describing the alkyl chain conformational order on the molecular level is thus crucial for the understanding of such materials function in their respective chromatographic applications. In this context, chromatographic measurements provide information about the properties of the bonded phase as a function of mobile phase, temperature, stationary phase support, solute hydrophobicity and polarity. Such studies, however, offer only indirect information about the bonded phase morphology.

In the present work, FT IR and solid-state NMR spectroscopy were employed, since both methods can be used to get complementary information about the influence of various external parameters – such as surface coverage, solvents, pressure, sample temperature, different synthetic pathways and solid supports – on the alkyl chain conformation and dynamics in chromatographic column material. The conformational order was examined primarily by variable temperature FT IR spectroscopy, where several conformation sensitive vibration bands can be analysed. Symmetric and anti-symmetric CH_2 and CD_2 stretching modes were utilized for getting qualitative information about the changes in the conformational order as a function of the aforementioned parameters. Quantitative information about the amounts of various gauche conformers (integral values over the whole chains) was obtained by the analysis of the CH_2 wagging band region between 1330 and 1400 cm^{-1}. The analysis of the CD_2 rocking bands in selectively deuterated alkyl chains provided the amount of gauche conformers at a specific (deuterated) methylene segment. The silane functionality and degree of cross-linking of the silane ligands on the silica surface were evaluated by ^{29}Si CP/MAS NMR spectroscopy, while the conformational order and mobility of the alkyl chains were studied by ^{13}C CP/MAS NMR spectroscopy.

At first, the influence of solid supports on the conformational order of C_{30} SAMs, prepared on zirconia, titania and two different silica gels (ProntoSil and LiChrospher) was examined. Here, we have chosen C_{30}-alkyl modified systems because of their remarkable shape selectivity properties. Variable temperature FT IR studies were carried out to assess the conformational properties of C_{30} SAMs over a broad temperature range. From the analysis of both the CH_2 stretching and CH_2 wagging bands it was found that alkyl chain flexibility and conformational disorder increase with increasing sample temperature. The highest degree of conformational order exists for SAMs prepared on titania, followed by zirconia and the silica supports. It was demonstrated that the observed temperature dependence of the conformational order is dominated by the changes in the number of kink/gtg conformers. On the basis of these data it can be summarized that for the C_{30} SAMs the solid substrate plays an important role for the actual conformational state of the attached alkyl chains.

Owing to such important role of the solid support for the conformational order, the studies were extended to C_{18}-alkyl modified systems which are frequently applied in chromatographic separations. The effect of solid supports and surface coverage on the conformational order of alkyl chains of commercially available carbon clad zirconia based supports (DiamondBond$^{®}$-C_{18}) and synthesized C_{18}-alkyl modified silica based supports was probed in the dry state using variable temperature FT IR and solid-state ^{13}C NMR spectroscopy. From FT IR spectroscopy, the conformational order of alkyl chains tethered to the substrates was examined by the analysis of CH_2 symmetric and anti-symmetric stretching bands. An analysis of the CH_2 wagging band data was not possible for the present DiamondBond$^{®}$-C_{18} systems due to overlap with other peaks, which are likely due to the phenyl-graphite layer modified substrate. The temperature dependence of the ^{13}C NMR and FT IR data was found to be consistent. It was observed that carbon clad zirconia based substrates exhibit a higher conformational order of alkyl chains and high thermal stability, which should be the major molecular reason for the enhanced performance of the DiamondBond®-C_{18} systems during chromatographic separations.

The conformational behavior of non-deuterated and selectively deuterated C_9, C_{18} and C_{22} chains (deuterated positions: carbons C-4, C-6 and C-12) alkyl modified silica gels in the dry state was examined by variable temperature FT IR spectroscopy. For these studies, the desired information was obtained by the analysis of CH_2 wagging, CD_2 stretching and CD_2 rocking bands. It was found that the conformational order critically depends on the actual alkyl chain length, chain position and sample temperature. Particular emphasis was given to the impact of the external pressure employed during sample preparation on the alkyl chain conformations. Two series of samples were examined: (i) samples which were prepared at normal pressure (method I), and (ii) samples which experienced a pressure of about 10 kbar during preparation (method II). It was observed that the samples prepared via method II, which experienced an external pressure, are characterized by a lower amount of gauche conformers. This substantial increase of conformational order was attributed to better alkyl chain packing along with a gain of intermolecular chain interactions.

Furthermore, the influence of surface coverage on C_{18}-alkyl modified silica gels with surface coverages varying from 2 to 8.2 $\mu mol/m^2$, was examined. The degree of cross-linking of the fixed alkylsilanes was examined by ^{29}Si NMR spectroscopy. The highest degree of cross-linking among the present samples resulted for C_{18}-alkyl modified silica gels with the highest surface coverage of 8.2 $\mu mol/m^2$. The influence of surface coverage on the alkyl chain conformational order was studied on a qualitative basis by ^{13}C NMR spectroscopy (chemical shifts of inner methylene segments), and the CH_2 symmetric and anti-symmetric stretching (band position and band width) and CH_2 wagging bands. Both spectroscopic methods provided consistent results, i.e., the conformational order decreases with increasing sample temperature and increases with higher surface coverage. Even for the highest surface coverage, a finite number of gauche defects was derived, and wagging band progressions, which are characteristic for all-trans chains, were not observed. In addition, the influence of the alkyl chain conformational order on chromatographic shape recognition was assessed through measurements of selectivity factors for shape selective solute probes. In agreement with the FT IR and NMR results, the selectivity factors ($\alpha_{TBN/BaP}$) generally decrease with increasing surface coverage. These data provide strong evidence for the relationship between chromatographic performance (i.e., shape recognition) and stationary phase conformational order.

The influence of the synthetic routes on the conformational order of C_{18} and C_{30}-alkyl modified silica gels was studied via samples obtained from the solution and surface polymerization routes. Variable temperature FT IR and ^{13}C CP/MAS NMR measurements were performed in order to get complementary information about the molecular properties of the alkyl chains. Again, as deduced from the spectral changes of the NMR and FT IR measurements, the conformational order of the tethered alkyl chains experiences an influence by the actual synthetic route. Despite the similar surface coverage, an increase in conformational order of the alkyl modified silica gels was observed for the C_{18} surface polymerized phases, which indicates that the alkyl chains are in the more ordered conformational state with fewer gauche contents and more trans bonds. Again, it was proved that surface coverage plays a major role on the conformational order of the alkyl modified systems. In this context, it was concluded that the influence of surface coverage dominates over the influence of alkyl chain lengths on the conformational order of stationary phase materials.

Finally, the impact of various solvents on the alkyl chain conformation was examined by FT IR spectroscopy. It was found that the C_{18} phases and mobile phase interact significantly, resulting in pronounced changes of the alkyl chain conformational order. Furthermore, it was observed that some deuterated solvents exhibit isotope effects, which again is reflected by a different alkyl chain conformational order as compared to the corresponding stationary phases with nondeuterated solvents.

In summary, it could be demonstrated that the present FT IR and NMR techniques are powerful tools for the evaluation of the alkyl chain conformational properties in stationary phase materials On the basis of the present investigations it can be stated that the solid substrates, surface coverage, pressure, synthetic routes, solvents and sample temperature play an important role in determining the conformational order and mobility of the alkyl chains in the stationary phase materials, which in turn determine their chromatographic performance. However, these materials are not yet completely understood on a molecular level. A lot more work is necessary, comprising different experimental techniques. In particular, this holds for the examination of such systems under "real" chromatographic conditions, i.e., including solvent, analyte and by considering pressure effects.

Chapter 7

Zusammenfassung

Die Umkehrphasen-Flüsigkeitschromatographie (RP-HPLC) stellt eine wichtige Methode zur Trennung und Analyse chemischer Verbindungen dar. Die entsprechenden Stationärphasen sind dadurch charakterisiert, dass an deren Oberflächen lange Alkylketten chemisch fixiert sind. Man geht davon aus, dass die Effizienz und Selektivität während des Chromatographie-Trennvorgangs durch die konformative Ordnung und die Dynamik der gebundenen Alkylketten wesentlich bestimmt werden. Studien in jüngerer Zeit haben gezeigt, dass die konformative Ordnung und die dynamischen Eigenschaften der Alkylketten sowie der damit verbundene Einfluss auf die Retention und Selektivität dieser Materialien beim chromatographischen Prozess von verschiedenen externen Parametern wie Alkylkettenlänge, Oberflächenbedeckung oder Temperatur abhängen. Für das Verständnis der Funktionsweise dieser Systeme ist es deshalb sehr wichtig, die Ordnung der fixierten Alkylketten auf molekularer Ebene im Detail zu kennen. In diesem Zusammenhang liefern chromatographische Messungen wichtige Aussagen über die Eigenschaften der gebundenen Phase (d.h. die fixierten Alkylketten) in Abhängigkeit von der gewählten mobilen Phase und stationären Trägerphase, der Hydrophobie und Polarität des Lösungsmittels und der Temperatur. Allerdings sind damit nur indirekte Informationen über die Morphologie der chemisch gebundenen Phase zugänglich.

In der vorliegenden Arbeit wurden FT IR- und Festkörper-NMR-spektroskopische Methoden zur Charakterisierung chromatographischer Trennphasen – in der Hauptsache chemisch modifizierte Silikagele – eingesetzt, die einen direkten Zugang zum Einfluss verschiedener externer Parameter (wie Oberflächenbedeckungsgrad, Lösungsmittel, Probentemperatur,

Syntheseroute für die Stationärphase) auf die Konformation und Dynamik der gebundenen Alkylketten erlauben. Die konformative Ordnung wurde in erster Linie mit Hilfe temperaturabhängiger FT IR-Messungen untersucht, wobei hier mehrere konformationssensitive Schwingungsbanden zur Verfügung stehen. Symmetrische und antisymmetrische CH_2- und CD_2-Streckschwingungen liefern dabei qualitative Informationen über die Änderung der konformativen Ordnung als Funktion der oben genannten Parameter. Quantitative Informationen über die mittlere Anzahl an Gauche-Konformeren in den Alkylketten sind über die Analyse der CH_2-Kippschwingungen im Bereich von 1330 bis 1400 cm^{-1} zugänglich. Die Analyse der CD_2-Pendelschwingungen in selektiv deuterierten Alkylketten liefert die Anzahl an Gauche-Konformeren an einem spezifisch deuterierten Methylensegment. Die Silanfunktionalität und der Vernetzungsgrad der Silanliganden an der SiO_2-Oberfläche wurden mit Hilfe der ^{29}Si CP/MAS-NMR-Spektroskopie bestimmt. Zusätzliche Aussagen über die Konformationsordnung und Mobilität der Alkylketten wurden aus ^{13}C CP/MAS-Untersuchungen erhalten.

Zunächst wurde der Einfluss des Trägermaterials der stationären Phase auf die konformative Ordnung von C_{30} SAMs (Self-Assembled-Monolayers) auf den Trägern ZrO_2, TiO_2 und zwei verschiedenen Silikagelen (ProntoSil und LiChrospher) untersucht. Die C_{30}-Trennphase wurde insbesondere deshalb gewählt, weil sie sich durch ein ausgezeichnetes formselektives Trennverhalten auszeichnet. Es wurden temperaturabhängige FT IR-Untersuchungen durchgeführt, um die konformativen Eigenschaften der C_{30}-SAMs über einen großen Temperaturbereich zu studieren. Die Analyse der CH_2-Streck- und CH_2-Kippschwingungsbanden ergab, dass die Flexibilität und die konformative Unordnung mit steigender Probentemperatur zunehmen. Die größte konformative Ordnung wurde für die auf TiO_2 gebundenen SAMs gefunden, gefolgt von den Proben mit ZrO_2- und SiO_2-Trägern. Es konnte gezeigt werden, dass die beobachtete Temperaturabhängigkeit der konformativen Ordnung maßgeblich durch die Änderung der Anzahl an kink/gtg-Konformeren bestimmt wird. Mit Hilfe dieser Untersuchungen ließ sich belegen, dass das für die hier untersuchten C_{30}-SAMs gewählte Trägermaterial eine wesentliche Rolle für den Konformationszustand der gebundenen Alkylketten spielt.

Um dieser wichtigen Rolle des Trägermaterials weiter nachzugehen, wurden die Untersuchungen auf C_{18}-Trennphasen ausgeweitet, die bei chromatographischen Trennverfahren sehr häufig zum Einsatz kommen. In diesem Zusammenhang wurde mit Hilfe von temperaturabhängigen FT IR- und ^{13}C Festkörper-NMR-Messungen der Einfluss des stationären Trägermaterials und des Oberflächenbedeckungsgrads auf die konformative Alkylkettenordnung bei einem kommerziell erhältlichen Material (DiamondBond$^{®}$-C$_{18}$) und bei einem selbst synthetisierten C$_{18}$-modifizierten Silikagel untersucht. Die IR-Untersuchungen basierten ausschließlich auf der Analyse der symmetrischen und antisymmetrischen CH$_2$-Streckschwingung. Die CH$_2$-Kippschwingungen konnten im Fall der DiamondBond$^{®}$-C$_{18}$ Systeme nicht analysiert werden, da andere Banden, die wahrscheinlich von Schwingungen der Phenyl-Graphitschicht herrühren, die CH$_2$-Kippschwingungen überdecken. Es zeigte sich, dass die Temperaturabhängigkeiten der ^{13}C NMR- und der FT IR-Daten sehr gut übereinstimmen. Das kommerzielle DiamondBond$^{®}$-C$_{18}$ System wies ferner eine deutlich höhere konformative Ordnung der Alkylketten und eine höhere thermische Stabilität auf. Es ist deshalb davon auszugehen, dass dies der wesentliche Grund für die gesteigerte Trennleistung der kommerziellen DiamondBond$^{®}$-C$_{18}$ Systeme im chromatographischen Prozess darstellt.

Temperaturabhängige FT IR-Untersuchungen wurden ferner an undeuterierten und deuterierten (deuteriert an den Positionen C-4, C-6 und C-12) C$_9$-, C$_{18}$- und C$_{22}$-Ketten in alkylmodifizierten Silikagelen vorgenommen, wobei CH$_2$-Kippschwingungen, CD$_2$-Streckschwingungen und CD$_2$-Pendelschwingungen analysiert wurden. Es wurde beobachtet, dass die konformative Kettenordnung stark von der Kettenlänge, der Position innerhalb der Kette und von der Probentemperatur abhängt. Ein besonderes Augenmerk galt dem Einfluss des bei der Probenvorbereitung angelegten Drucks auf die Alkylkettenkonformation. In diesem Zusammenhang wurden zwei Probenserien untersucht: (i) Proben, die bei Normaldruck präpariert wurden (Methode I), und (ii) Proben, bei denen während der Probenvorbereitung ein Druck von 10 kbar ausgeübt wurde (Methode II). Dabei wiesen die mit Methode II präparierten Proben eine niedrigere Anzahl an Gauche-Konformeren auf. Diese Erhöhung der konformativen Ordnung wurde auf eine effektivere Packung der Alkylketten und den damit verbundenen stärkeren zwischenmolekularen Wechselwirkungen der Ketten zurückgeführt.

Weiterhin wurde der Einfluss des Oberflächenbedeckungsgrads in C$_{18}$-alkylmodifizierten Silikagelen (Bedeckungsgrad im Bereich von 2 bis 8.2 µmol/m^2) untersucht. Der Vernetzungsgrad der gebundenen Alkylsilane wurde mit Hilfe der ^{29}Si NMR-Spektroskopie bestimmt. Der höchste Vernetzungsgrad wurde bei den C$_{18}$-alkylmodifizierten Silikagele mit der höchsten Oberflächenbedeckung von 8.2 µmol/m^2 beobachtet. Die konformative Ordnung der Alkylketten wurde mit Hilfe der ^{13}C NMR-Spektroskopie (chemische Verschiebung der Methylensegmente) sowie symmetrischen und antisymmetrischen CH$_2$-Streckschwingungen (Bandenposition und -breite) und CH$_2$-Kippschwingungen ermittelt. Die verschiedenen Untersuchungsmethoden liefern übereinstimmende Ergebnisse, wobei die Konformationsordnung mit steigender Bedeckung zu- und mit zunehmender Temperatur abnimmt. Selbst bei der höchsten Oberfächenbedeckung liegt noch eine endliche Anzahl an Gauche-Defekten vor. Kippschwingungs-Progressionen, wie sie in Gegenwart von all-trans-Ketten auftreten, wurden deshalb nicht beobachtet. Weiterhin wurde der Einfluss der Alkylketten-Konformation auf die chromatographische Formselektivität überprüft. In Übereinstimmung mit den FT IR- und den NMR-Daten, nimmt der chromatographische Selektivitätsfaktor mit zunehmender Oberflächenbedeckung ab. Diese Resultate zeigen eindeutig, dass ein direkter Zusammenhang zwischen chromatographischer Trennleistung (z.B. Formerkennung) und der konformativen Ordnung der Alkylketten der stationären Phase besteht.

Zwei verschiedene Syntheseverfahren, die sog. Lösungspolymerisation und die sog. Oberflächenpolymerisation, wurden angewandt, um deren Einfluss auf die Konformationsordnung in C$_{18}$- und C$_{30}$-modifizierten Silikagelen aufzuklären. Dabei kamen wiederum temperaturabhängige FT IR- und ^{13}C CP/MAS NMR-Messungen zum Einsatz. Die beobachteten spektralen Veränderungen sowohl bei den NMR- als auch bei den FT IR-Messungen zeigen deutlich, dass die Konformationsordnung der gebundenen Alkylketten auch vom jeweils gewählten Syntheseansatz abhängt. Trotz vergleichbarer Oberflächenbedeckungen wurde eine höhere Konformationsordnung bei der oberflächenpolymerisierten C$_{18}$-Phase beobachtet. Die Alkylketten befinden sich also in einer geordneten Phase mit weniger Gauche-Anteilen und mehr Trans-Bindungen. Die Möglichkeit verschiedene Konformationszustände auszubilden verringert sich mit längerer Alkylkette und bei niedrigeren Temperaturen. Auch bei diesen Untersuchungen wurde festgestellt, dass die Oberflächenbedeckung eine wichtige Rolle für die konformative Ordnung in alkylmodifizierten Silikagelen spielt. In diesem Zusammenhang ließ sich nachweisen, dass der Oberflächenbedeckungsgrad für die konformative Kettenordnung der stationären Phase wichtiger ist als die Alkylkettenlänge.

Im letzten Abschnitt der Arbeit wurde mit Hilfe der FT IR-Spektroskopie der Einfluss des Lösungsmittels auf die Alkylkettenkonformation untersucht. Die Analyse der CH_2-Streckschwingungen der untersuchten C_{18}-Phasen weist auf eine ausgeprägte Wechselwirkung zwischen der mobilen Phase und der stationären Phase hin, was sich in einer gegenüber dem lösungsmittelfreien System veränderten konformativen Kettenordnung widerspiegelt. Für einige deuterierte Lösungsmittel wurde ein Isotopeneffekt beobachtet, der sich in einer gegenüber dem entsprechenden undeuterierten Lösungsmittel veränderten Konformationsordnung niederschlägt.

Die hier durchgeführten FT IR- und Festkörper-NMR-Messungen konnten zeigen, dass diese spektroskopischen Verfahren zur Klärung der Konformationseigenschaften von chromatographischen Trennphasen bestens geeignet sind. Auf der Basis der hier erzielten Ergebnisse lässt sich zusammenfassend feststellen, dass die feste Trägermatrix, der Oberflächenbedeckungsgrad, der Druck, die gewählte Syntheseroute, das Lösungsmittel und die Probentemperatur wichtige Kenngrößen darstellen und die Konformationsordnung und die Beweglichkeit der gebundenen Alkylketten von chromatographischen Trennphasen in entscheidender Weise beeinflussen, was wiederum Auswirkungen auf das chromatographische Trennverhalten hat. Für ein umfassendes Verständnis der molekularen Vorgänge beim chromatographischen Trennprozess sind jedoch noch weitere Untersuchungen unter Berücksichtigung verschiedener, insbesondere auch spektroskopischer Messverfahren notwendig. Dies schließt vorzugsweise Experimente unter „chromatographischen" Bedingungen, d.h. die Gegenwart von Elutionsmittel und externem Druck, ein.

Abbreviations

RP	Reversed Phase
HPLC	High Performance Liquid Chromatography
FT IR	Fourier Transform Infrared Spectroscopy
NMR	Nuclear Magnetic Resonance
CP/MAS	Cross Polarisation Magic Angle Spinning
SAMs	Self Assembled Monolayers
PAH	Polycyclic Aromatic Hydrocarbons
α_{RP}	Surface Coverage
eg	End gauche
gg	Double gauche
gtg	Gauche-trans-gauche
gtg'	Kink
ppm	parts per million
RIS	Rotational Isomeric State
S_{BET}	Specific surface area according to Brunnett, Elliott and Teller
SRM	Standard Reference Material
$\alpha_{TBN/BaP}$	Shape selectivity factor
TBN	Tetrabenzonapthalene
BaP	Benzo[a]pyrene
HDZr-C_{18}	High Density DiamondBond$^{®}$-C_{18} phases

LDZr-C_{18} Low Density DiamondBond®-C_{18} phases

Si-C_{18} C_{18}-alkyl modified silica gels

C_{18} Surface C_{18} Surface polymerized phases

C_{18} Solution C_{18} Solution polymerized phases

C_{30} Surface C_{30} Surface polymerized phases

C_{30} Solution C_{30} Solution polymerized phases

List of Figures

117

List of Tables

References

1. Vansant, E. F.; VanDerVoort P.; Vrancken, K. C. *Characterization and Chemical Modification of the Silica Surface*, Elsevier, Amsterdam, 1995.

2. Neue, U. D. *Encylopedia of Analytical Chemistry*, Meyers, R. A (ed), Wiley, Chichester, 2000.

3. Doyle, C. A.; Dorsey, J. G. *The Handbook of Liquid Chromatography*, Katz, E.; Eksteen, R.; Schoenmakers, P.; Miller, N. (Eds.), Marcel Dekker, NewYork, 1998, p.293.

4. Sander, L. C.; Lippa, K. A.; Wise, S. A. *Anal. Bioanal. Chem.* **2005**, *382*, 646.

5. Saito, Y.; Ohta, H.; Jinno, K. *J. Sep. Sci.* **2003**, *26*, 225.

6. Sentell, K. B. *J. Chromatogr.* **1993**, *656*, 231.

7. Nawrocki, J.; Dunlap, C. J.; Carr, P. W.; Blackwell, J. A. *Biotechnol. Prog.* **1994**, *10*, 561.

8. Pesek, J. J.; Matyska, M. T. *Encylopedia of Chromatography,* Cases, J. (Eds.), Marcel Dekker, NewYork, 2001.

9. Przybyciel, M.; Majors, R. E. *LCGC* **2002**, *20(6)*, 516.

10. Kurganov, A.; Trudinger, U.; Isaeva, T.; Unger, K. K. *Chromatographia* **1996**, *42*, 217.

11. McNeff, C.; Zigan, L.; Johnson, K.; Carr, P. W.; Wang, A.; Weber-Main, A. M. *LCGC* **2000**, *18*, 514.

12. Ellwanger, A.; Matyska, M. T.; Albert, K.; Pesek, J. J. *Chromatographia* **1999**, *49*, 4244.

13. Sentell, K. B. Henderson, A.N. *Anal. Chim. Acta* **1991**, *246*, 139.

14. Sentell, K. B.; Dorsey, J. G. *J. Chromatogr.* **1989**, *461*, 193.

15. Sander, L. C.; Wise, S. A. *Adv. Chromatogr.* **1986**, *25*, 139.

16. Sander L. C.; Wise S. A. *Anal. Chem.* **1995**, *67*, 3284.

17. Pursch, M.; Sander, L. C.; Albert, K. *Anal. Chem.* **1996**, *68*, 4107.

18. Tanaka, N.; Tokuda, Y.; Iwaguchi, K.; Araki, M. *J. Chromatogr.* **1982**, *239*, 761.

19. Pursch, M.; Sander, L. C.; Albert, K. *Anal. Chem. News & Features* **1999**, *71*, 733A.

20. Pursch, M.; Vanderhart D. L.; Sander L. C.; Gu, X.; Nguyen. T.; Wise S. A.; Gajewski, D. A. *J. Am. Chem. Soc.* **2000**, *122*, 6997.

21. Neumann-Singh, S.; Villanueva-Garibay, J.; Müller, K. *J. Phys. Chem. B* **2004**, *108*, 1906.

22. Pursch, M.; Sander, L. C.; Egelhaaf, H. J.; Raitza, M.; Wise, S. A.; Oelkrug, D.; Albert, K. *J. Am. Chem. Soc.* **1999**, *121*, 3201.

23. Singh, S.; Wegmann, J.; Albert, K.; Müller, K. *J. Phys. Chem. B* **2002**, *106*, 878.

24. Sander, L. C.; Callis, J. B.; Field, L. R. *Anal. Chem.* **1983**, *55*, 1068.

25. Doyle, C. A.; Vickers, T. J.; Mann, C. K.; Dorsey, J. G. *J. Chromatogr. A* **2000**, *877*, 25.

26. Doyle, C. A.; Vickers, T. J.; Mann, C. K.; Dorsey, J. G. *J. Chromatogr. A* **2000**, *877*, 41.

27. Pemberton, J. E.; Ho, M.; Orendorff, C. J.; Ducey, M. W. *J. Chromatogr. A* **2001**, *913*, 243.

28. Ducey, M. W.; Orendorff, C. J.; Pemberton, J. E. *Anal. Chem.* **2002**, *74*, 5576.

29. Klatte, S. J.; Beck, T. L.; *J. Phys. Chem.* **1993**, *97*, 5727.

30. Klatte, S. J.; Beck, T. L.; *J. Phys. Chem.* **1995**, *99*, 16024.

31. Yarovsky, I.; Aguilar, M. I.; Hearn, M. T. W. *J. Chromatogr. A* **1994**, *660*, 75.

32. Yarovsky, I.; Aguilar, M. I.; Hearn, M. T. W. Anal. Chem. **1995**, *67*, 2145.

33. Lochmüller, C. H.; Hunnicutt, M. L.; Mullaney, J. F. *J. Phys. Chem.* **1985**, *89*, 5770.

34. Jinno, K.; Ibuki, T.; Tanaka, N.; Okamoto, M.; Fetzer, J. C.; Biggs, W. R.; Griffiths, P. R.; Olinger, J. M. *J. Chromatogr.* **1989**, *461*, 209.

35. Jinno, K.; Wu, J.; Ichikawa, M.; Takata, I. *Chromotographia* **1993**, *37(11-12)*, 627.

36. Lochmüller, C. H.; Marshall, S. F.; Wilder, D. R. *Anal. Chem.* **1980**, *52*, 19.

37. Snyder, R. G. *J. Chem. Phys.* **1967**, *47*, 1316.

38. Doyle, C. A.; Vickers, T. J.; Mann, C. K.; Dorsey, J. G. *J. Chromatogr. A* **1997**, *779*, 91.

39. Zeigler, R.C.; Maciel, G. E. *J. Phys. Chem.* **1991**, *95*, 7345.

40. Sindorf, D.W.; Maciel, G. E. *J. Am. Chem. Soc.* **1983**, *105,*1848.

41. Albert, K.; Evers, B.; Bayer, E. *J. Magn. Reson.* **1985**, *62*, 428.

42. Pursch, M.; Strohschein, S.; Handel, H.; Albert, K. *Anal. Chem.* **1996**, *68,* 386.

43. Gangoda, M.; Gilpin, R. K.; Figueirinhas, J. *J. Phys. Chem.* **1989**, *93*, 4815.

44. Zeigler, R. C.; Maciel, G. E. *J. Am. Chem. Soc.* **1991**, *113,* 6349.

45. Kelusky, E. C.; Fyfe, C. A. *J. Am. Chem. Soc.* **1986**, *108,*1746.

46. Ba, Y.; Chagolla, D. *J. Phys. Chem. B* **2002**, *106*, 5250.

47. Chagolla, D.; Ezedine, G.; Ba, Y. *Micropor. Mesopor. Mater.* **2003**, *64*, 155.

48. Mendelsohn, R.; Snyder, R. G. *Biological Membranes*, ed. Merz, Jr. K. M.; Roux B.; Birkhäuser (Eds.) Boston, 1996.

49. Layne, J. *J. Chromatogr. A* **2000**, *877*, 25.

50. Majors, R. E. *LCGC* **1997**, *15(11),* 1009.

51. Unger, K. K. *Porous Silica*, Elsevier, Amsterdam, 1979.

52. Unger, K. K. *Packings and Stationary Phases in Chromatographic Techniques*, Marcel Dekker, New York, 1998.

53. Sander L. C.; Wise S. A. *Anal. Chem.* **1984**, *56*, 504.

54. Nawrocki, J.; Rigney, M. P.; McCormick, A.; Carr, P. W. *J. Chromatogr. A* **1993**, *657*, 229.

55. Rimmer, C. A.; Sander, L. C.; Wise, S. A.; Dorsey, J. G. *J. Chromatogr. A* **2003**, *1007*, 11.

56. Pursch, M.; Brindle, R.; Ellwanger, A.; Sander, L. C.; Bell, C. M.; Handel, H.; Albert, K. *Solid State Nucl. Magn. Reson.* **1997**, *9*, 191.

57. Sander L. C.; Pursch, M.; Wise S. A. *Anal. Chem.* **1999**, *71*, 4821.

58. Grun, M.; Kurganov, A.; Schacht, S.; Schuth, F.; Unger, K. K. *J. Chromatogr. A* **1996**, *740*, 1.

59. Henry, R. A. *American Laboratory* **2002**, *34(22)*, 18.

60. Dorsey J. G.; Cooper, W. T. *Anal. Chem.* **1994**, *66*, A857.

61. Boehm, C.; Leveiller, F.; Moehwald, H.; Kjaer, K.; Als-Nielsen, J.; Weissnuch, I.; Leiserowitz, L. *Langmuir* **1994**, *10*, 830.

62. Bierbaum, K.; Grunze, M.; Baksi, A. A.; Chi, L. F.; Schrepp, W.; Fuchs, H. *Langmuir* **1995**, *11*, 2143.

63. Bierbaum, K.; Linzler, M.; Woell, C. H.; Grunze, M.; Haehner, G.; Heid, S.; Effenberger, F. *Langmuir* **1995**, *11*, 512.

64. Sander, L. C.; Sharpless, K. E.; Craft, N.; Wise, S. A. *Anal. Chem.* **1994**, *66*, 1667.

65. Emenhiser, C.; Sander, L. C.; Schwartz, S. J. *J. Chromatogr.* **1995**, *707*, 205.

66. Emenhiser, C.; Englert, G. E.; Sander, L. C.; Ludwig, B.; Schwartz, S. J. *J. Chromatogr. A,* **1996**, *719*, 333.

67. Martire, D. E.; Boehm, R. E. *J. Phys. Chem.* **1983**, *87*, 1045.

68. Ducey, M. W.; Orendorff, C. J.; Pemberton, J. E.; Sander, L. C. *Anal. Chem.* **2002**, *74*, 5585.

69. Ducey, M. W.; Orendorff, C. J.; Pemberton, J. E.; Sander, L. C. *Anal. Chem.* **2002**, *75*, 3360.

70. Horvath, C.; Melander, W.; Molnar, I. *J. Chromatogr.* **1976**, *125*, 129.

71. Karger, B. L.; Gant, J. R.; Hartkopf, A.; Weiner, P. H. *J. Chromatogr.* **1976**,*128*, 65.

72. Vailaya, A.; Horvath, C. *J. Phys. Chem. B* **1997**, *101*, 5875.

73. Vailaya, A.; Horvath, C. *J. Phys. Chem. B* **1998**, *102*, 701.

74. Marqusee, J. A.; Dill, K. A. *J. Chem. Phys.* **1986**, *85*, 434.

75. Dill, K. A. *J. Phys. Chem.* **1987**, *91*, 1980.

76. Dill, K. A.; Naghizadeh, J.; Marquesee, J. A. *Annu. Rev. Phys. Chem.* **1988**, *39*, 425.

77. Dorsey, J. G.; Dill, K. A. *Chem. Rev.* **1989**, *89*, 331.

78. Buszewski, B.; Jezierska, M.; Welnak, M.; Berek, D. *J. High. Resol. Chromatogr.* **1998**, *21*, 267.

79. Berendsen, G. E.; de Galan L. *J. Liq. Chromatogr.* **1978**, *1*, 561.

80. Snyder, R. G.; Strauss, H. L.; Elliger, C. A. *J. Phys. Chem.* **1982**, *86*, 5145.

81. Cameron, D. G.; Casal, H. L.; Mantsch, H. H.; Boulanger, Y.; Smith, I. C. P. *Biophys. J.* **1981**, *35*, 1.

82. Flory P. J. *Statistical Mechanics of Chain Molecules*, Wiley, New York, 1969.

83. Senak, L.; Davies, M. A.; Mendelsohn, R. *J. Phys. Chem.* **1991**, *95*, 2565.

84. Neumann-Singh, S. Ph. D Dissertation, Universität Stuttgart, 2003

85. Maciel, G. E.; Sindorf, D.W. *J. Am. Chem. Soc.* **1980**, *102,* 7606.

86. Tonelli, A.E. *Macromolecules* **1978**, *11*, 565.

87. Tonelli, A.E. *Spectroscopy and Polymer Microstructure, The Conformational Connection,* VCH Publishers, New York, 1989.

88. SRM 869a, Column Selectivity Test Mixture for Liquid Chromatography (Polycyclic Aromatic Hydrocarbons), Certificate of Analysis, NIST, Gaithersburg, MD, 1998.

89. Srinivasan, G.; Pursch, M.; Sander, L. C.; Müller, K. *Langmuir* **2004**, *20*, 1746.

90. Nuzzo, R. G.; Fusco, F. A.; Allara, D. L. *J. Am. Chem. Soc.* **1987**, *109*, 2358.

91. Porter, M. D.; Bright, T. B.; Allara, D. L.; Chidsey, C. E. D. *J. Am. Chem. Soc.* **1987**, *109*, 3559.

92. Fadeev, A. Y.; Helmy, R.; Marcinko, S. *Langmuir* **2002**, *18*, 7521.

93. Griffiths P. R.; de Haseth, J. A. *Fourier Transform Infrared Spectroscopy*, Wiley-Interscience, New York, 1986.

94. Maroncelli, M.; Qi, S. P.; Strauss, H. L.; Snyder, R. G. *J. Am. Chem. Soc.* **1982**, *104*, 6237.

95. Wolfangel, P.; Meyer, H. H.; Bornscheuer, U. T.; Müller, K. *Biochim. Biophys. Acta* **1999**, *1420*, 121.

96. Senak, L.; Moore D.; Mendelsohn, R. *J. Phys. Chem.* **1992**, *96*, 2749.

97. Chia, N.; Mendelsohn, R. *J. Phys. Chem.* **1992**, *96*, 10543.

98. Srinivasan, G.; Kyrlidis, A.; McNeff, C.; Müller, K. *J. Chromatogr. A* **2005**, *1081*, 132.

99. McNeff, C. V.; Stoll, D. R.; Carr, P. W.; Hawker, D. R.; Kyrlidis, A.; Gaudet, G. The Pittsburgh Conference on Analytical Chemistry and Applied Spectroscopy, New Orleans, Louisiana, 2001.

100. Carr, P. W.; McNeff, C. V.; Stoll, D. R.; Hawker, D. R.; Kyrlidis, A.; Gaudet, G. Eastern Analytical Symposium, Somerset, New Jersey, 2001.

101. Earl, W. L.; Vanderhart, D. L. *Macromolecules* **1979**, *12*, 762.

102. Geschke, D. *Z. Phys. Chem.* **1968**, *239*, 138.

103. Shibanuma, T.; Asada, H.; Ishi, S.; Matsui, T. *Jpn. J. Appl. Phys.* **1983**, *22*, 1656.

104. Tabony, J. *Prog. NMR Spectrosc.* **1980**, *14*, 1.

105. Myers, K. J. *Molecular Magnetism and Magnetic Resonance Spectroscopy*, Prentice Hall, Englewood Cliffs, NJ, 1973.

106. Perez, E.; Vanderhart, D. L. *J. Polym. Sci. B* **1987**, *25*,1637.

107. Srinivasan, G.; Neumann-Singh, S.; Müller, K. *J. Chromatogr. A* **2005**, *1074*, 31.

108. Ho, M.; Cai, M.; Pemberton, J. E. *Anal. Chem.* **1997**, *69*, 2613.

109. Maroncelli, M.; Strauss, H. L.; Snyder, R. G.; *J. Phys. Chem.* **1985**, *82*, 2811.

110. Maroncelli, M.; Strauss, H. L.; Snyder, R. G.; *J. Phys. Chem.* **1985**, *89,* 4390.

111. Hostetler, M. J.; Stokes, J. J.; Murray, R. W. *Langmuir* **1996**, *12*, 3604.

112. Voicu, R.; Badia, A.; Morin, F.; Lennox, R. B.; Ellis, T. H. *Chem. Mater.* **2001**, *13*, 2266.

113. Badia, A.; Demers, L.; Cuccia, L.; Lennox, R. B. *J. Am. Chem. Soc.* **1997**, *119*, 2682.

114. Badia, A.; Lennox, R. B.; Reven, L. *Acc. Chem. Res.* **2000**, *33*, 475.

115. Templeton, A. C.; Hostetler, M. J.; Kraft, C. T.; Murray, R. W. *J. Am. Chem. Soc.* **1998**, *120*, 1906.

116. Wong, P. T. T. *Biophys. J.* **1994**, *66*, 1505.

117. Siminovitch, D. J.; Wong, P. T. T.; Mantsch, H. H. *Biochemistry* **1987**, *26*, 3277.

118. Wong, P. T. T.; Mantsch, H. H.; Snyder, R. G. *J. Chem. Phys.* **1983**, *79*, 2369.

119. Wong, P. T. T.; Chagwedera, T. E.; Mantsch, H. H. *J. Chem. Phys.* **1987**, *87*, 4487.

120. Wong, P. T. T.; Chagwedera, T. E.; Mantsch, H. H. *J. Mol. Str.* **1991**, *247*, 31.

121. Srinivasan, G.; Sander, L. C.; Müller, K. (Manuscript in preparation)

122. Badia, A.; Gao, W.; Singh, S.; Demers, L.; Cuccia, L.; Reven, L. *Langmuir* **1996**, *12*, 1262.

123. Albert, K.; Schmid, J.; Pfleiderer, B.; Bayer, E. *Chemically Modified Surfaces*, Mottola, H. A.; Steinmetz, J. R. (eds) Elsevier, Amsterdam, 1992, pp 105-116.

124. Cheng, J.; Fone, M.; Ellsworth, M. W. *Solid State Nucl. Magn. Reson.* **1996**, *7*, 135.

125. Kamlet, M. J.; Abboud, J. M.; Abraham, M. H.; Taft, R. W. J. Org. Chem. **1983**, *48*, 2877.

126. Kimata, K.; Hosoya, K.; Araki, T.; Tanka, N. *Anal. Chem.* **1997**, *69,* 2610.

127. Baweja, R. *Anal. Chim. Acta,* **1987**, *192*, 345.

128. Lochmüller, C. H.; Hunnicutt, M. L. *J. Phys. Chem.* **1986**, *90,* 4318.

129. Gandoda, M. E.; Gilpin, R. K. *Langmuir* **1990**, *6*, 941.

130. Sander, L. C.; Glinka, C. J.; Wise, S. A. _Anal. Chem._ **1990**, _62,_1099.

131. Kazakevich, Y. V.; Lobrutto, R.; Chan, F.; Patel, T. _J. Chromatogr. A_ **2001**, _913,_75.

132. Sentell, K. B.; Dorsey, J. G. _Anal. Chem._ **1989**, _61,_930.

Curriculum Vitae

Personal:

Name   : Gokulakrishnan Srinivasan
Date of Birth : 17th December 1978
Nationality   : Indian

Academic:

Since October 2001 — **Ph.D.**

Institute for Physical Chemistry, University of Stuttgart, Germany under the guidance of Prof. Dr. Klaus Müller.

Thesis: Solid-State NMR and FT IR Studies on Chromatographic Column Materials

1999 – 2001 — **M.Sc.**

M.Sc. in Chemistry at Bharathiar University, Coimbatore, Tamil Nadu, India.

Project completed under the guidance of Dr. R. Murugesan at the Regional Sophisticated Instrumentation Centre, **Indian Institute of Technology - Madras (IIT), Chennai, India.**

Thesis: Characterisation of Gallstones in Southern India

1996 – 1999 — **B.Sc.**

B.Sc. in Chemistry at Bharathiar University, Coimbatore, Tamil Nadu, India.

Publications:

1. **G. Srinivasan,** C. Meyer, K. Albert, K. Müller, Influence of synthetic routes on the conformational order of n-alkyl modified silica gels. Journal of Chromatography A (Manuscript submitted).

2. **G. Srinivasan,** L. C. Sander K. Müller, Influence of surface coverage on the conformation and mobility of C_{18}-modified silica gels, Analytical and Bioanalytical Chemistry (in press).

3. **G. Srinivasan**, A. Kyrlidis, C. McNeff, K. Müller, Investigation on conformational order and mobility of DiamondBond-C18 and C18-alkyl modified silica gels by FTIR and Solid-state NMR spectroscopy, Journal of Chromatography A 2005, 1081, 132.

4. L. Armelao, H. Bertagnolli, S. Gross, V. Krishnan, U. Lavrencic-Stangar, K. Müller, B. Orel, **G. Srinivasan**, E. Tondello, A. Zattin, Zirconium and hafnium oxoclusters as molecular building blocks for the preparation of nanostructured silica based inorganic-organic hybrid materials. Journal of Materials Chemistry 2005, 15, 1954.

5. **G. Srinivasan**, S. Neumann-Singh, K. Müller, Conformational order of n-alkyl modified silica gels as evaluated by Fourier transform infrared spectroscopy, Journal of Chromatography A 2005, 1074, 31.

6. **G. Srinivasan**, M. Pursch, L. C. Sander, K. Müller, FTIR studies of C_{30} self-assembled monolayers on silica, titania, and zirconia, Langmuir 2004, 20, 1746.

7. **S. Gokulakrishnan**, M. Ashok and V. Jayanthi, Analysis of gallstone - Critical appraisal of various techniques, Gastroenterology Today 2001, 5, 145.

8. **S. Gokulakrishnan**, R. Murugesan, S. Mathew, R. Prashanthi, A. C. Ashok, H. Ramesh, G. Sivakumar, R. Surendran, V. Jayanthi, Predicting the compositions of gallstones by infrared spectroscopy, Tropical Gastroenterology 2001, 22, 87.